# THE HUMAN SKELETON

# THE BASICS

*The Human Skeleton: The Basics* provides an accessible overview of the basic characteristics of the human skeleton that can be used to understand the life and (sometimes) death of persons represented only by their skeletons. Topics covered include:

- Osteology (bones) and Odontology (teeth),
- Estimating Ancestry, Sex, and Age at Death,
- Calculating Stature,
- Skeletal Anomalies,
- Cultural Modifications,
- Cranial and Dental Pathological Conditions,
- Disease and Trauma.

*The Human Skeleton: The Basics* is an essential read for students, faculty, and professionals in anthropology, biology, forensics, and criminal justice.

**Steven N. Byers** has a Ph.D. in Anthropology from the University of New Mexico (UNM), USA. Now retired, he worked for a number of years on various campuses of UNM, and currently serves on the Anthropology Consensus Body and Odontology Consensus Body of the American Academy of Forensic Sciences Standards Board.

# The Basics Series

*The Basics* is a highly successful series of accessible guidebooks which provide an overview of the fundamental principles of a subject area in a jargon-free and undaunting format.

Intended for students approaching a subject for the first time, the books both introduce the essentials of a subject and provide an ideal springboard for further study. With over 50 titles spanning subjects from artificial intelligence (AI) to women's studies, The Basics are an ideal starting point for students seeking to understand a subject area.

Each text comes with recommendations for further study and gradually introduces the complexities and nuances within a subject.

**ANTHROPOLOGY OF REPRODUCTION**
*SALLIE HAN AND CECÍLIA TOMORI*

**SCIENCE COMMUNICATION**
*MASSIMIANO BUCCHI AND BRIAN TRENCH*

**PROJECTION DESIGN**
*DAVIN E. GADDY*

**ETHNOGRAPHY**
*SUSAN WARDELL*

**BAYESIAN STATISTICS**
*THOMAS J. FAULKENBERRY*

**FAT STUDIES**
*MAY FRIEDMAN*

**PROPAGANDA**
*NATHAN CRICK*

**THE HUMAN SKELETON**
*STEVEN N. BYERS*

For more information about this series, please visit: www.routledge.com/The-Basics/book-series/B

# THE HUMAN SKELETON

# THE BASICS

Steven N. Byers

Routledge
Taylor & Francis Group

NEW YORK AND LONDON

Designed cover image: spxChrome, Getty Images

Note on figures: Unless otherwise specified, all drawings are by the author or from drawings in the public domain.

First published 2026
by Routledge
605 Third Avenue, New York, NY 10158

and by Routledge
4 Park Square, Milton Park, Abingdon, Oxon OX14 4RN

*Routledge is an imprint of the Taylor & Francis Group, an informa business*

ISBN: 978-1-032-78454-0 (hbk)
ISBN: 978-1-032-77261-5 (pbk)
ISBN: 978-1-003-48794-4 (ebk)

DOI: 10.4324/9781003487944

Typeset in Bembo
by Taylor & Francis Books

# CONTENTS

# FIGURES

# TABLES

# INTRODUCTION

When talking about human bone, the well-known anthropologist, Clyde Collins Snow, often stated that "bones made good witnesses, never lying, never forgetting, and that a skeleton, no matter how old, could sketch the tale of a human life, revealing how it had been lived, how long it had lasted, what traumas it had endured and especially how it had ended." This line from Snow's obituary (New York Times, May 16, 2014) is an appropriate starting quote for this book since it is about what can be revealed about the life, and sometimes death, of persons represented only by their skeletons.

The study of the human skeleton is termed human skeletal biology or human osteology. This subject requires a thorough knowledge of human bones and teeth, their features, and the different forms they can take in different populations, different sexes, and in persons of different ages. During an osteological analysis, the bones must be recognized as human, nonhuman, or prehuman (fossil ancestor). If human, it must be determined if all bones are present or if some of the 206 bones are missing or if there is more than one person represented in the collection. Each bone must be able to be identified, placed in the proper position within the skeleton, and if paired, the side (left or right) the bone came from must be able to be determined. This knowledge must be so extensive that even fragments, sometimes very small fragments, can be recognized and identified. The same applies to the study of the teeth (dentition), termed odontology. Teeth must be recognized as human or nonhuman, the permanent dentition must be distinguished from the deciduous (milk) teeth, and the number, type (incisor, canine, premolar, molar), side, and placement (upper or lower jaw) must be determined. As with bones, even

DOI: 10.4324/9781003487944-1

fragments of these tissues must be recognized. When a high level of knowledge is reached, bones and teeth can reveal many features of the life of persons represented only by osteological and odontological remains.

## WHAT BONES AND TEETH REVEAL

A good way to state what bones and teeth reveal is through a series of questions that can be asked when a skeleton is being analyzed. Of the large number of questions, there are seven that are most commonly asked.

1   Are these the bones of a human, prehuman, or nonhuman?
2   What is this person's ancestral group?
3   Was this person male or female?
4   Do the bones and teeth indicate at what age this person died?
5   How tall was the person?
6   Do the bones and teeth show any diseases (pathological conditions), including trauma, which may have negatively affected this person in life and/or helped cause the person's death?
7   Is there any indication in the skeleton as to what activities this person engaged in during life, like squatting to prepare food (e.g., grinding grain) or using teeth to hold objects (nails when repairing shoes)?
8   Are there any odd features of the bones and teeth that do not occur in most skeletons?

Of these eight, questions 2 thru 5 represent what is called the biological profile of the person, and it is these four (they could be called the 'big four') that virtually all people who analyze human skeletons (or early human ancestors) will want to answer. Knowing the ancestry, sex, age at death, and stature helps in understanding the answers to other questions.

Although the eight given above are the most common, there are numerous other questions that can, and have been, asked. Questions such as: what is the chemical composition of bone, including isotopes, that can help identify what the person ate while growing up? How does bone grow, and how does that help understand the life of persons? How does bone heal itself after an injury, including fracture? How did individual bones evolve, and why are they shaped the way they are? How does bone react to stress, and how does it change shape (remodel) throughout life? What is the internal structure of bone, both visible (macroscopic) and invisible (microscopic)? Are there indications in the skeleton as to the place of origin of the person or the person's ancestors? What function do bones perform and how does that affect the way they look (i.e., how does form influence function)? Which traits seen in bone are due to genetics, and which are due to the environment (e.g., nutrition, stress)? If the bones are from a prehuman, are they in the line of human evolution or in a side branch that went extinct or evolved into some other form? And many others.

The answers to these questions have multiple uses. If the bones and teeth being analyzed are part of a police (forensic) investigation, the answers (especially to the 'big' four) are used to help identify a person from a missing persons file and (if present) provide information on cause and manner of death. This helps to bring justice to victims of murder, genocide, and terror attacks as well as help identify persons killed by natural catastrophes (e.g., hurricanes, tsunamis, floods) or human disasters (e.g., airplane crashes).

Another use of these answers is to reveal the way past people lived. By gathering together (aggregating) the answers to the above (and other) questions, information (statistics) can be developed on a population represented by skeletal material (e.g., percentage of skeletons that are female and male to calculate the sex ratio). This type of research usually involves studying the remains from prehistoric and historic (formal and informal) cemeteries, where dozens, if not hundreds of skeletons are analyzed. Aggregated demographics (ancestry, sex, age at death) provide information on population origin, population growth, population age structure, life expectancy, and other similar information. Aggregated pathological data can provide information on dietary deficiencies, or disease loads. Other aggregate data may illuminate how they practiced medicine (i.e., treated diseases) or engaged in warfare. This type of analysis focuses on learning as much as possible about a past (extinct) population from the remains of their dead.

If diseases of bones are of interest, they will be identified and categorized, and examples of especially interesting pathological conditions will be studied and reported on. From there, aggregate pathological information from multiple skeletons can be used to reveal the origin of diseases and their spread through time and/or between towns, countries and even continents. This may also include research on how sex differences in disease frequency may infer cultural traditions and even gender roles. These and other studies related to pathological conditions use the presence of disease to reach the same goal of understanding the population represented by the skeletons of their people.

Lastly, if the bones are from pre-humans or early humans, their evolutionary significance will be explored. Using information on bone form and function, attempts will be made to determine if the individual walked upright, was able to use their hands to manipulate objects (e.g., use tools), or other information on behavior. The form of the bones will also be compared with early ancestors (if known) and later individuals to help with placing them in (or out of) the line of human evolution. When enough skeletons of the same type (e.g., *Homo erectus*, Neanderthals) are uncovered, the goal will be the same as that discussed above: that of understanding how the population lived.

## SKELETAL ANALYSIS METHODS

When analyzing a skeleton two main techniques are used: visual examination (also called qualitative assessment) and quantitative metric measurement (also called osteometry/craniometry). Visual examination involves using the eye, both unaided or magnified, to answer the eight (and other) questions. With extensive knowledge of the human skeleton, features that help estimate ancestry, sex, age at death, pathological conditions, and anomalies can be recognized. A simple example of this method is the observation that females, on the average, are smaller and less robust than males (studies have shown they average around 92% the size of males). Using this knowledge when visually examining two skeletons of different sizes, it would be reasonable to say that the larger is more likely male and the smaller more likely female. Similarly, if the larger skeleton has larger and more rugged muscle markings (i.e., rough areas where muscles attach to bone through tendons), the estimation that it is from a male is further supported. This extends to complex observations that require considerable prior knowledge of skeletal variations to untangle the life of the person represented by osteological remains.

Osteometry is the quantitative measurement of human bones using the Imperial system (inches, feet, pounds) or the Metric system (millimeters, centimeters, kilograms). Postcranial bones (those below the head) are measured for lengths, widths, thicknesses, and sometimes angles. The results of these measurements can be used for a variety of reasons, including estimating sex and calculating the living height (stature) of persons. Craniometry is that branch of osteometry that involves measurements of the skull: heights, breadths (widths), lengths, angles, and arcs. Craniometric measurements also can be used to estimate sex as well as ancestry and other traits. These measurements require several types of calipers and a specially designed instrument, called an osteometric board or bone board, to measure postcranial bones.

In addition to visual and metric assessment, many other methods are available when analyzing skeletons. X-raying bones can reveal their inner structure, such as the unusual thinness of the bones of the braincase of an ancient Egyptian named Khety who lived during the XIIth Dynasty. This thinning is a rare condition of unknown cause seen in fewer than 1% of modern people. X-rays can also show the presence of foreign objects, such as a bullet found in the vertebra of a murder victim that was not easily seen with the unaided eye. More complete pictures of the inner structure of bones can be seen using CT-scans as when its use on a 250,000- to 350,000-year-old archaic human skull revealed what is believed to be the first known example of a meningioma, a type of tumor of the membrane (meninges) surrounding the brain (and other parts of the nervous system). Chemical analysis of bone also can reveal features of the life of persons. The levels of isotopes of carbon in Native American bones from late prehistoric Ontario helped to show the shift from occasional consumption of maize (an early type of corn) to more frequent consumption of this plant, possibly indicating the shift from hunting and gathering to maize agriculture. The ratio of other isotopes found in bone of recently deceased persons can help identify the region where they grew up by comparing their ratios with the known ratios in various areas of a geographic region. This helped to show that 6 of 36 formerly enslaved persons from a revolutionary war cemetery in Charleston,

South Carolina, were born in Africa and were presumably brought to North America as slaves, rather than being the children of enslaved persons. The characteristics of newly fractured bones have been studied by experimentally breaking fresh human or nonhuman bone in the controlled environment of a laboratory. These experiments help to distinguish bones that were broken around the time of death (and could be the cause of death) from those that occurred after death (so-called taphonomic changes). Scanning electron microscopy (SEM) has also been used, such as when grooves on the front teeth of some prehistoric Native Americans showed fine scratches that indicate a material, such as animal sinew for bow strings or plant strips for basketry, was repeatedly passed back-and-forth over their edges. The analysis of DNA extracted from bone and teeth can be used to help identify ancestry and sex. This is just a sample of the less commonly used methods when analyzing human bones.

Once the data has been gathered, it is analyzed using a variety of other methods to help answer the questions posed above. These analytic methods often do not answer the questions definitely, but more often provide a probable answer. This is particularly true for the 'big four.' A skeleton is not said to be definitely that of a person of European ancestry but is 'probably' of that ancestral group. Similarly, a skeleton is not said to be male or female, rather it is given a probability that it is male or female. For age at death and stature, a range of values is usually provided (e.g., death occurred in the early twenties, the person was between 5 ft. 6 in. and 5 ft. 10 in. tall), rather than a single number (e.g., the person was 23 years old at death, the person was 5 ft. 8 in. tall). These probabilities and ranges of values are based on research using osteological collections of human skeletons where ancestral group, sex, age at death, and stature (among other things) are known. Using these collections in the United States, the frequency of traits that distinguish Euro-Americans (Whites) from Afro-Americans (Blacks) from Asian Americans (Asians), or males from females have been calculated, and these frequencies can then be used to estimate group membership of a skeleton. Similarly, the ranges of ages for changes in bone that occur during growth have been derived from studies of the skeletons of known age, as have the relationship between living height and bone lengths.

How data from skeletons of known ancestry, sex, age and stature are used is best illustrated with an example. Looking at the heads men and women reveals that males are more likely to have mounds, called supra-orbital ridges, on their lower forehead right above their eyes, while females have smooth foreheads above the eyes. These supra-orbital ridges range from absent to fairly large elevations that are easily seen and felt. Research has shown that it is best to score these on a five-point scale from absent (category=1) to very large (category=5) with three intermediate categories. Table 1.1 presents the results of one such study of 447 skulls of known sex and their counts within each of the five categories. Notice that, of the 101 skulls with smooth foreheads and no supra-orbital ridges, 15 are male and 86 are female, indicating that smooth foreheads are more common in women than men. These values can be converted into frequencies by dividing the number of male and female skulls by the total (i.e., 15/101=.149, 86/101=.851) as seen in the right two columns of the table. Again, the higher frequency of females without supra-orbital ridges is shown. These frequencies can also be thought of as probabilities (designated with the letter 'p'); that is, if a skull is found with no supra-orbital ridges, then there is a p=.851 (85.1% chance) that it is female and p=.149 (14.9% chance) that it is of a male. This is the situation with most visual and measured traits; the assignment to a particular group (in this case, male or female) is not hard-and-fast. Rather, it is best given as a probability and therefore is an estimate of group membership. In the early days of skeletal analysis, group membership was usually stated as absolute, but current research shows it is better stated in terms of probability.

Several other items in Table 1.1 need to be examined further. Notice that the probability of male and female is less convincing in the middle three categories. This is typical of many of the visually examined traits; that is, the expressions of the characteristic at ends of the scale (in this case, no supra-orbital ridges versus the largest expression of the trait) have the highest probability of group membership (in this case, p=.851 and p=.957 for female and male, respectively), while the middle category probabilities are less different. Another interesting feature of this table is the small number of skulls with the largest supra-orbital ridges. This propensity for few individuals to be in the extreme category is also common

*Table 1.1* Frequency of Skulls with Supra-orbital Ridges

| Category | Description | Number of Skeletons | | | Frequency | |
|---|---|---|---|---|---|---|
| | | Total | Male | Female | Male | Female |
| 1 | Smooth, No Supra-orbital Ridges | 101 | 15 | 86 | 0.149 | 0.851 |
| 2 | Barely Visible | 134 | 52 | 82 | 0.388 | 0.612 |
| 3 | Visible but Small | 120 | 84 | 36 | 0.700 | 0.300 |
| 4 | Larger, Easily Seen | 69 | 57 | 12 | 0.826 | 0.174 |
| 5 | Very Large, Very Noticeable | 23 | 22 | 1 | 0.957 | 0.043 |

Source: Modified from Table 4 of Walker (2008).

among the visual traits; conversely, usually the smallest expression or the absence of a trait are seen in many skeletons. The last property to consider is how uneven the categories are in relation to which are most likely to be male or female. When looking at a characteristic (in this case, supra-orbital ridges), it is expected that two would most likely be female, two male, and the middle one either/or (indeterminate). However, the situation in Table 1.1 is fairly typical of all visual traits scored on at 5-point (or other) scale. The data do not fit into neat, preconceived arrangements like the one just mentioned; rather it is more like what is seen in Table 1.1 where categories 3, 4, and 5 are most likely male and only 1 and 2 are likely female. Human skeletal data is fairly messy and does not fit easily into exact patterns.

A similar situation exists with metric measurements. A study of the width of the top end (called the proximal epiphysis) of 339 shin bones (tibias) of Euro-Americans shows that those widths that are over 74 millimeters are most likely male while those under that amount are female. This is expected since, as stated earlier, males are larger on the average than females. However, the study showed that the correct classification into male or female was not 100%, rather 90% of tibias were properly sexed using this criterion with 10% being estimated as the wrong sex. As done with visual traits, these percents are easily converted into probabilities by dividing by 100; thus, there is a probability of $p=.90$ (90% chance) that tibias above 74 mm are male and those below that size are female, with a p=.10 probability (10% chance) of improper classification. As with the above discussion of supra-orbital ridges, the imprecise nature of group assignment using skeletal remains is shown.

Although the above two examples show how a single trait can be used to estimate group membership, most skeletons have several traits that can provide that estimate. Since each of these traits are expressed as frequencies/probabilities (derived from the studies of skeletal collections of known biological profile), they can be combined to arrive at a best overall probability. There are a number of methods for doing this, but the simplest is to add all of them together within each group and divide by the number of traits scored to arrive at a (very) rough estimate. As an example, in addition to the supraorbital ridges, there are three other traits that help to distinguish male from female skulls (these are illustrated in Figure 4.2 in Chapter 4). For the purposes of illustration, suppose the probabilities of the four traits of the skull being male are:.700,.880,.957,.628, and female are:.300,.120,.043,.372. The overall probability would be (.700+.880+.957+.628) ÷ 4 =.791 for male, and (.300+.120+.043+.372) ÷ 4 =.209 for female. These probabilities indicate that the best estimate of sex is male. There is software, both freely available on the internet and for sale by various agencies, that combine probabilities using more complex methods than the simple averaging example just given.

There are a number of other methods commonly used to analyze data from human osteology that will be discussed in the following chapters. Most use methods explained in introductory and advanced statistics classes and textbooks. One of the most commonly employed of these methods is discriminant function analysis (DFA) which uses multiple metric measurements to estimate group membership (i.e., ancestral group, sex). Computer programs calculate a single number from these measurements and assign group membership based on whether it is above or below another number, called a sectioning point, that divides the groups in question. Single and multiple regression analysis is another commonly used complex method that uses metric measurements to calculate another unknown measurement; the most common is long bone lengths used to estimate the living height (stature) of an individual. These, and other methods, are too complex to be discussed in a book of this nature, but interested readers can consult statistics textbooks for further information.

## PURPOSE OF BOOK

The purpose of this book is to present the reader with the basics of the study of the human skeleton, and the simplest observations and metric measurements used to answer the eight questions posed above. In addition, there will be brief discussions of the more complex methods, with each chapter providing a list of books and articles that will help the reader become more versed in this subject. First, the reader will be presented with a chapter that describes human bones and teeth, and their features, needed to understand the contents of subsequent material. The next chapters present methods used to estimate answers to the 'big four' questions (ancestry, sex, age at death, stature), followed by chapters that describe the most common anomalies and pathological conditions (diseases) seen in human skeletons as well as common cultural modifications. The next chapter will discuss those supplementary methods used when identifying persons in a forensic investigation, when determining how populations lived, when analyzing diseases, and when researching early humans and prehumans that provide answers to questions posed in skeletal analysis.

Needless to say, the reader will not be an expert human skeletal biologist after reading this book. Such expertise is only gained by intense study over a lifetime of viewing osteological remains. However, the reader will understand what can be told (or, more appropriately, 'estimated') about the life and (sometimes) death of persons represented only by their skeleton.

## SUMMARY

1   The human skeleton has a number of characteristics that provide information on how a person lived and (sometimes) how they died.
2   The study of the human skeleton is called human skeletal biology or human osteology.
3   There are eight questions about a person represented only on their skeletal remains that are most often asked.

4   The four most important of these questions involve the biological profile of persons: ancestry, sex, age at death, and stature.

5   Other important questions involve osteological indications of a person's activities in life, diseases they may have suffered from, and skeletal indicators of relationships with other persons.

6   Visual examination and metric measurements are the two basic methods used to help reveal the biological profile, and other features, of a person's life.

## FURTHER READING

The best book on human skeletal biology is TD White, MT Black, and PA Folkens, *Human Osteology*, 3rd ed. (Cambridge, MA: Academic Press, 2012). Others include WM Bass, *Human Osteology: A Laboratory and Field Manual*, 5th ed. (Springfield, MO: Missouri Archaeological Society, 2005), and DG Steele and CA Bramblett, *The Anatomy and Biology of the Human Skeleton* (College Station, TX: Texas A&M University, 1988).

Clyde Snow's Obituary by RD McFadden, Clyde Snow, Sleuth Who Read Bones From King Tut's to Kennedy's, Dies at 86 (New York Times, 2014). https://www.nytimes.com/2014/05/17/us/clyde-snow-forensic-detective-who-found-clues-in-bones-dies-at-86.html

Table 1.1 has been modified from Table 4 of PL Walker Sexing skulls using discriminant function analysis of visually assessed traits (*American Journal of Physical Anthropology*, 136(1):39–50, 2008).

# BASICS OF HUMAN OSTEOLOGY AND ODONTOLOGY

As mentioned in Chapter 1, the analysis of human skeletal remains requires a person to have a thorough knowledge of the human skeleton and dentition. Skeletal biologists must be able to recognize all bones and teeth of the skeleton, even from small pieces, and where they fit within the body. Because of their complexity, entire books are written about the skeleton and dentition. However, there are a few topics that stand out as most important to know when doing basic skeletal analysis. These topics typically use only whole bones and teeth, rather than fragments, and it is these topics that will be described in the most detail in this chapter.

The purpose of this chapter is to provide a brief overview of human bones and teeth so that readers will understand the subjects described in the remaining chapters of this book. After that description, the components seen in many bones and teeth will be described along with a discussion of their internal structure.

## HUMAN OSTEOLOGY

Although a variety of abnormal conditions can add to the number, the adult human skeleton ordinarily is composed of 206 bones. These range in size from small, angular polygons (e.g., the bones of the wrist) to long, heavy limb bones with straight or gently curving shafts. Each bone in the body is given a name as well as a side designation (if paired). Thus, the bone of the upper arm is called the humerus, and persons normally are born with one on the right and one on the left side. Some bones and skeletal structures are actually made up of a number of smaller bones; for example, the

DOI: 10.4324/9781003487944-2

wrist is composed of eight separate bones, while the skull consists of 22, plus three small bones (called ossicles) in each ear. Finally, with the exception of only one (the hyoid suspended below the mouth), all bones connect directly to (articulate with) at least one (and in many cases more than one) other bone.

Unfortunately, the word 'bone' can cause confusion since it has two different meanings in skeletal biology. The first meaning is a named bone of the skeleton. Each of the 206 bones has a name, such as humerus for the upper arm bone, or name and number, such as the $5^{th}$ lumbar vertebra of the vertebral (spinal) column. This is the meaning that most people give to the word 'bone.' Another use of the word is the material that the named bones are made of. Bone used in this context refers to the calcium, phosphate, and other minerals as well as cells (called osteons) that make up a named bone. To further complicate things, there is fibrous bone (that is made up of stiff but disorganized collagen fibers) and lamellar bone (the hard, dense, and well-organized bone that makes up the normal skeleton). Additionally, there is cortical bone which is the term for the surface bone of a named bone that is normally made up of lamellar bone. This contrasts to trabecular bone found inside named bones that has the appearance of a sponge but also is made up of lamellar bone. In addition, the word 'boney' is used when describing a feature of a named bone, such as 'the mastoid process of the skull is a boney knob that points downward from behind the ear.' (One way to distinguish these two meanings is the presence or absence of 'a' and 'the', as in the sentence "The surface of the shin bone is made up of lamellar bone.") Readers should remain aware of this dual use of the word to better understand the subjects described in this book.

Another note on words will help in the reading of this book. There are a number of features that appear in, and on, many bones of the skeleton. Some have similar sounding names and can have similar or different functions. For example, a tubercle is a small and rounded bump where tendons attach muscles to bone, but a tuberosity is a larger version of this bump that has a rough surface and is also a place where tendons attach. Similarly, many bones have condyles and epicondyles. A condyle is a part of a bone that helps form a joint. It can be rounded (as in the condyles of the thigh bone at the knee) or relatively flat (as in the top of the shin bone at the knee). It also can be a smooth boney mound on the surface of a bone. An epicondyle is a small bump near a condyle where muscles attach. There are two features that are similar in appearance but are distinct in at least one characteristic. A fossa (plural: fossas) is a pit that is broader and longer than it is deep while a foramen (plural: foramina) is a hole in, and usually through, a bone. The former has a boney bottom while the latter does not. Another feature are facets that appear in many parts of the skeleton and are small, smooth, and relatively flat areas that make up a joint. The bones of the spinal column (vertebrae) all have two pairs of facets, one pair on the top (to attach to the vertebra or feature above it) and one on the bottom (to attach to the vertebra or feature below it). A sulcus is a groove in bone; nerves or vessels (veins, arteries) run along the bottom of these grooves where they are protected from pressure from overlying muscles or outside forces. A trochanter is a rugged, boney bulge where muscles attach; they usually jut from the bone more than tubercles, tuberosities, and epicondyles. Lastly, many bones have processes; these are boney fingers that jut up from the surrounding surface and take many different forms.

When describing bones and features of the skeleton, it is helpful to think of it as a standing person with their arms at their side with their fingers extended and their palms facing forward. This is called the anatomical position. With this position in mind, one can think of bones being up toward the top or down toward the bottom of the body, from the upper or lower parts of the body, from the front or back of the skeleton, or from the (right or left) side. Unfortunately, sometimes these commonly used terms (up-down, upper-lower, top-bottom, front-back, or right-left) do not always make sense when describing the bones and features of the skeleton (this will become obvious throughout this book), so the study of the human body (human anatomy) has devised a series of terms called cardinal directions and planes, which can be used in any discussion of the skeleton. These terms are presented in Table 2.1 along with their definitions and more commonly used words.

Figure 2.1 presents a line drawing of the human skeleton showing the names of the major bones and how they join (articulate) with each other. For ease of study, the skeleton can be divided into two areas: cranial and postcranial. The cranial skeleton

*Table 2.1* Cardinal Directions and Planes Used in Describing the Skeleton

| Cardinal Direction | Description |
|---|---|
| Superior (cranial) | Up; point or region lying above another point or region |
| Inferior (caudal) | Down; point or region lying below another point or region |
| Medial | Central, point or region lying closest to the midline of the body |
| Lateral | Side, point or region lying away from the midline of the body |
| Anterior (ventral) | Front; point or region lying closest to the front of the body |
| Posterior (dorsal) | Back; point or region lying closest to the back of the body |
| Proximal | Point closest to an articular point with the body |
| Distal | Point farthest from an articular point with the body |
| Anatomical Planes | |
| Sagittal | Midline; plane cutting through the body that divides it into left and right halves |
| Coronal | Plane, at right angles to the midline (sagittal plane), that divides the body into front (anterior) and back (posterior) halves |
| Transverse | Plane that divides the body at the waist into upper (superior) and lower (inferior) sections |

(skull) is made up of the bones of the head and is often studied as a separate unit because of its complexity. The postcranium (all bones below the skull) is also divided for ease of study into two segments: the axial skeleton (those bones making up the vertebral column and rib cage) and the appendicular skeleton (made up of the bones of the shoulder, arms, hands, pelvis, legs, and feet). Learning the important features of each of these segments and the bones within them makes it easier to understand the topics in all other chapters of this book.

CRANIAL SKELETON

The skull is made up of 22 outwardly visible bones and three small bones (called ossicles) in each ear that cannot be seen easily. Except for the lower jaw and the ossicles, these articulate tightly at suture lines so that there is little movement between neighboring bones.

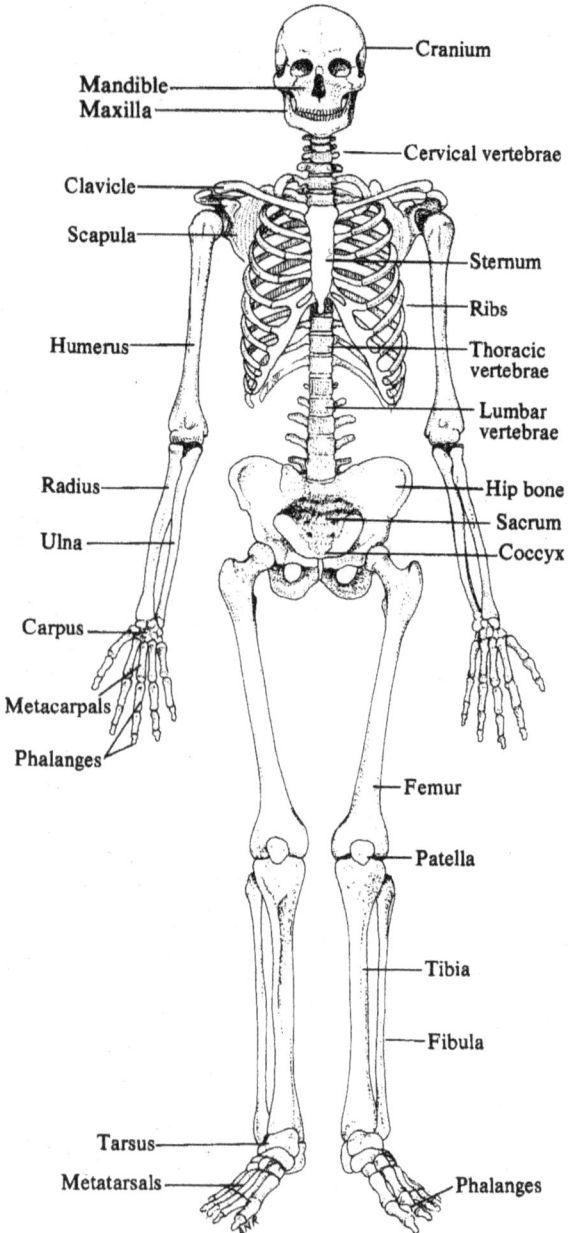

*Figure 2.1* Bones of the Human Skeleton
Source: From Figure 2.1 of *Digging Up Bones* by DR Brothwell, (1981) Ithaca, NY: Cornell University Press. Used with permission.

Because this structure holds many of the most important tissues needed for living (eyes, ears, nose, mouth, teeth) as well as other features like extra bones (also called ossicles), it is the most widely studied structure in the skeleton. To organize this study, the description of the cranium is divided into four topical areas. The first gives the name of the bones that make up the cranial skeleton and the various processes and foramina visible on their outside surface. The second deals with the sutures, their names and which bones they separate, and the third topic involves what are called landmarks on the cranium. Figures 2.2, 2.3, 2.4, and 2.5A illustrate these three features. The fourth and final skull area describes the various sinuses within bones of the skull; these deserve considera-tion because of their value in developing positive identifications of individuals (see Chapter 10 "Specialty Methods").

*BONES*

Figure 2.2 shows the skull from the front (anterior). Starting from the top (superiorly), the frontal is the forehead bone that makes up the front part of the braincase and the upper part of the eye orbits. As described in Chapter 1, there are thickened areas above each eye orbit called the superciliary arch but more commonly called the browridge that are most commonly found in males. Joining (articulating) with the frontal along the bottom (inferiorly) are the right and left zygomatic bones (plural: zygomatics); these "cheek bones" also make up the side (lateral) walls of the eye orbits. Forming the floor of each of the orbits is the maxilla (plural: maxillae); this paired bone also makes up the upper jaw, which holds the upper teeth. The maxillae join (articulate) laterally with the zygomatics and, in the middle (medially) at the top (super-iorly), with a pair of bones that form the bridge of the nose, called the nasal bones (plural: nasals). Below the nasals, the maxillae border the nose opening (nasal aperture) until they meet in the midline just above the teeth. Inside the nose are turbinal bones called nasal conchae (singular: concha) that supply the structure for the soft tissues involved in the sense of smell. To the rear of the nose is a thin vertical bone, called the vomer, which divides the back (posterior) nasal opening roughly in half. Finally, the lower jaw (mandible) contacts the maxillae along the rows of teeth (the

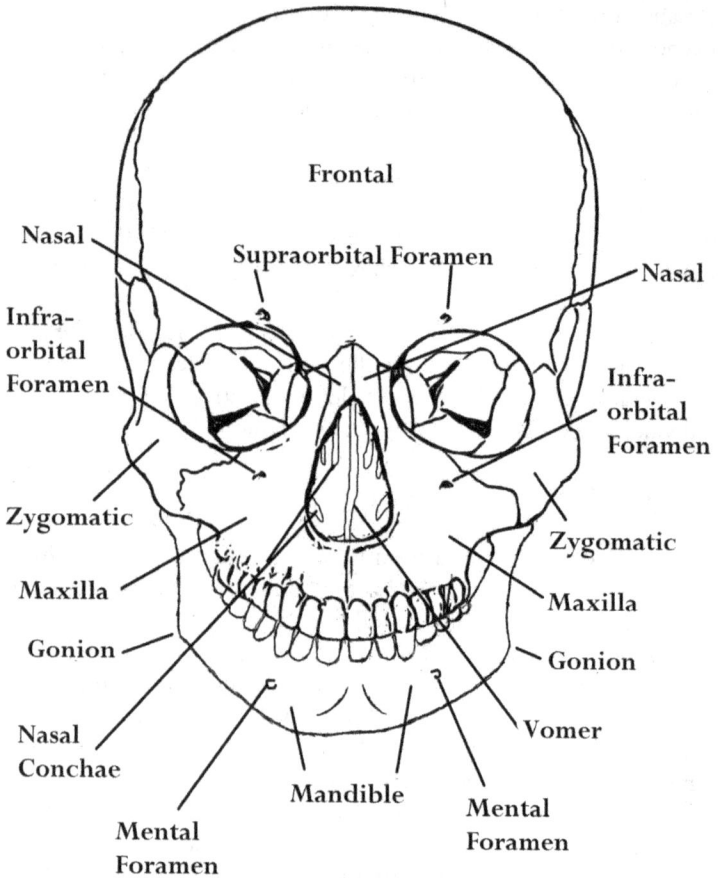

*Figure 2.2* Front View of the Skull Showing Bones, Foramina, and Landmarks

dental arcade); in this view, its main features are the mental foramina on either side of the midline. In addition, the inside of the
corners where the horizontal and vertical parts of the mandible
meet are tiny ridges of bone (spicules) called the pterygoid tuberosities (not pictured). Muscles attach to these that help pull the
lower jaw up and forward.

Figure 2.3 shows the side (lateral) view of the skull, where other
bones of both the face (called the splanchnocranium) and braincase

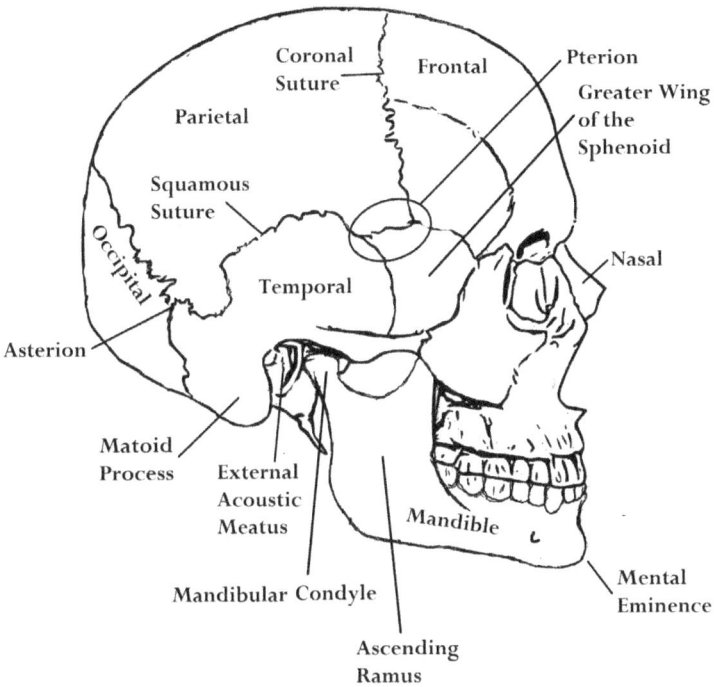

*Figure 2.3* Side View of the Skull Showing Bones, Features, Foramina, and Landmarks

(called the neurocranium) are visible. Starting with the front, the side of the frontal is seen as the anterior part of the braincase. Joining (articulating) with the frontal along its rear border is the parietal bones (plural: parietals); these paired (right and left) bones make up the middle of the braincase. At the back (posterior) of the parietals is the occipital that forms the back part of the braincase, and it curves under the back (posterior) half of the neurocranium to form much of the base of the skull. The temporal bone (plural: temporals) forms the lower part of the sides of the braincase; the most important features of these paired bones are the ear opening (called the external acoustic meatus) and the mastoid process (a boney knob that sticks downward). The temporal is also the bone where the mandible connects to the skull by the mandibular condyles of the lower jaw (the temporal-mandibular joint or TMJ). Finally, the temporals join anteriorly with an extremely complex bone called the sphenoid; the greater wing of this bone is located between the temporal and the zygomatics and connects (articulates) superiorly with the frontal and parietals. Two last features in this view are the mental eminence (also called the mental protuberance), which is a jutting of the chin near the lower (inferior) margin of the mandible in the midline, and the ascending ramus, which is the vertical part of this bone that contains the mandibular condyles. Not labelled in Figure 2.3 is the horizontal ramus, which is the horizontal part of the mandible.

The top (superior) and back (posterior) views of the skull are shown in Figure 2.4. The superior view (Figure 2.4A) shows the main bones of the braincase: frontal, right and left parietals, and occipital. The back (posterior) view of the skull (Figure 2.4B) also shows both parietals as well as the occipital. The main part of this latter bone is called the squamous (not labeled) and forms the entire rear of the skull.

The inferior view (Figure 2.5A) shows the occipital, inferior parts of the temporals, and lower part of the sphenoid. Important features of this view include the mandibular fossas (sockets for the condyles of the mandible on the temporal bones), the foramen magnum through the occipital (which allows for passage of the spinal cord into the brain), the occipital condyles (for attachment to the vertebral column), zygomatic arches (made up of the posterior part of the zygomatics) and the palate (mainly made up of the inferior maxillae). The back of the occipital is called the nuchal area and has several lines where the neck muscles attach, the upper one of which is called the superior nuchal line. This can become so large as to form ridges called nuchal crests (pictured in Figure 4.2), which in turn can produce a boney downward-pointing knob of bone in the midline called the external occipital protuberance, or sometimes called the inion hook. The nuchal lines are typical of many areas of muscle attachment (collectively called muscle markings) and are usually more rugged in males than females.

Within the skull bones are a number of foramina where nerves, veins, and arteries pass through. The supraorbital foramen (see Figure 2.2) is in the lower part of the frontal above each eye orbit. On the maxillae are infraorbital foramen (see Figure 2.2) below the

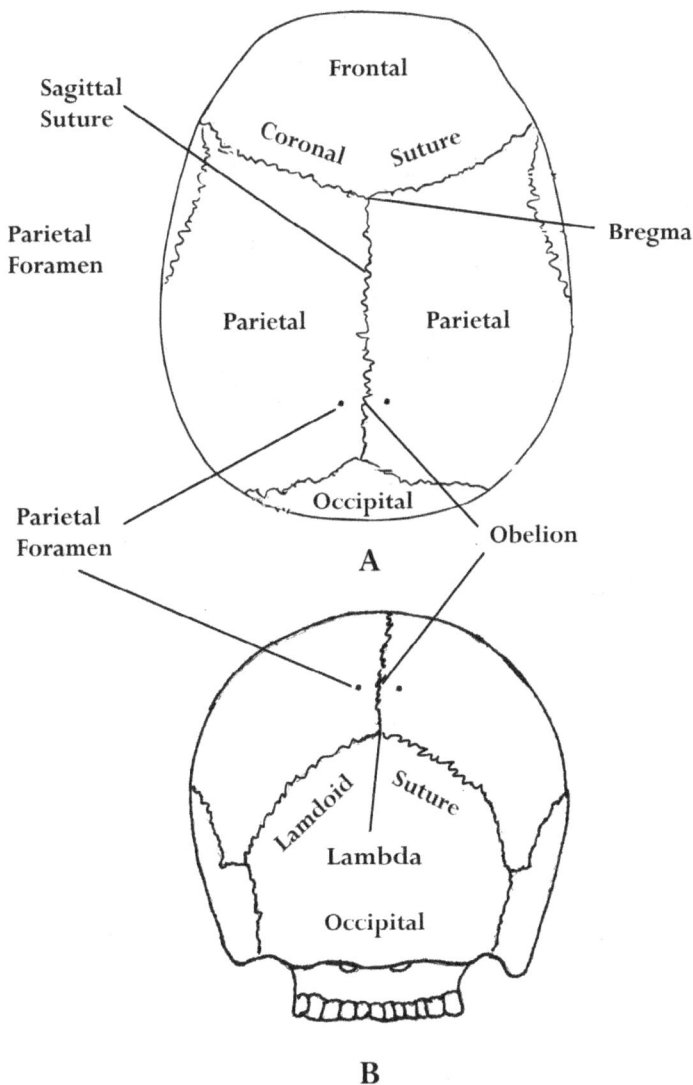

*Figure 2.4* Top View (A) and Rear View (B) of the Skull Showing Bones, Foramina, and Landmarks

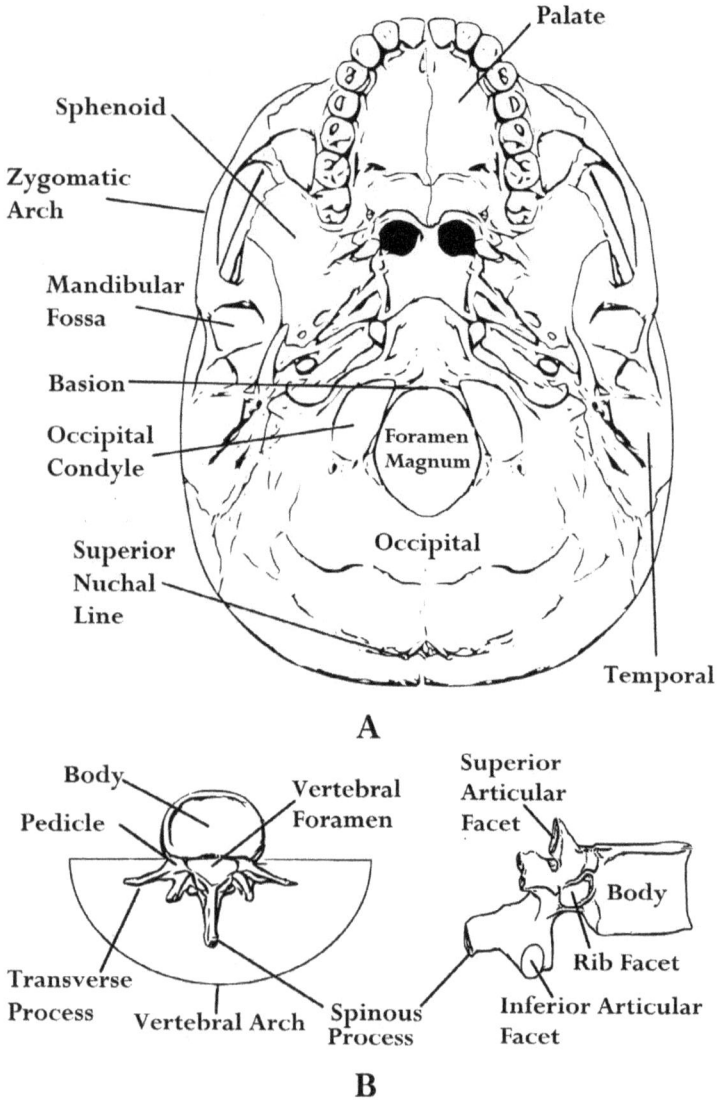

*Figure 2.5* Bottom View of the Skull (A) and Anatomy of a Vertebra (B)

eye orbits and the mental foramen (see Figure 2.2) are located on the lower jaw (mandible). The zygomatic bones also have for-amina, called the zygomaticofacial foramen (not pictured in Figure 2.2, but see Figure 7.1D). Each parietal usually, but not always, has the parietal foramen (see Figure 2.4A and B) toward the back and near the sagittal suture for the passage of a vein (the parietal emis-sary vein).

*SUTURES*

As just described, the skull is formed from a number of different bones joined with each other by special joints that appear as lines cut into the cranium. These lines are referred to as sutures. Understanding of the sutures is important, because most of these structures close with time and may become completely obliterated in old age. Thus, this feature can help estimate the age at death of individuals from their skeletons. The majority of the sutures are named for the bones that they separate. Thus, the internasal suture separates the nasal bones, and the zygomaxillary suture separates the zygomatics and maxillae. However, a number of sutures are not named in this manner. The coronal suture, visible in Figures 2.3 and 2.4A, separates the frontal and parietals; it ends at the point where the frontal meets the greater wing of the sphenoid. The sagittal suture, also visible in Figure 2.4A, separates the right and left parietal bones; it starts at the coronal suture and ends where the parietals meet the occipital. The lambdoid suture separates the parietals from the occipital bone (see Figure 2.4B); it curves across the back of the skull, ending where the occipital and parietals meet the temporals. The final, specially named suture separates the squama (wide, flat part) of the temporals from the parietals (see Figure 2.3); this is the squamous suture.

*LANDMARKS*

Over the years, it has been convenient to designate points on the skull for use in description or in measurement. There are many of these landmarks on the skull, but only those referred to in chapters of this book are illustrated in the figures. Bregma is the point where the sagittal suture ends anteriorly at the coronal suture in the sagittal plane (Figure 2.4A); basion (Figure 2.5A) is the most inferior point on the anterior border of the foramen magnum. On the rear of the skull is the point where the sagittal and lambdoid suture meet; this point, called lambda (see Figures 2.4B), also is in the midline. Pterion is not an actual point; rather it is the region where the greater wing of the sphenoid meets the frontal, parietal, and temporal (Figure 2.3). Asterion is the point where the squamosal suture ends at the lambdoid suture (Figure 2.3). Finally, gonion is the point where the ascending ramus meets the horizontal ramus (Figure 2.2).

## SINUSES

The final structures relevant to the study of the human skull are the pockets of air, called sinuses, within sections of some of the cranial bones. The most important of these are located in the frontal bone but the temporals, sphenoid, and maxillae all have these pockets. The frontal sinus is a complex open area that lies both above the upper border of the eye orbits and in the lower portion of the frontal in the midline (i.e., above nasal bones). The maxillary sinuses are large, uncomplicated areas in the upper jaw that appear to lighten these bones without sacrificing structural integrity. The sinuses of the other bones are more complex in nature since they are composed of multiple cells that are not easily seen even in x-ray. Research into sinuses, especially in the frontal, has indicated that their shape is unique enough in individuals that they can be used to positively identify a skeleton when antemortem radiographs are available for comparison (see Chapter 10 "Specialty Methods").

## AXIAL SKELETON

This part of the postcranium is composed of the sternum and the bones of the vertebral column (spinal column) and rib cage. The sternum (Figure 2.7A) is the breastbone that serves as an anchor for the anterior ends of the ribs. At the top of this bone is a roughly triangular part called the manubrium that joins (articulates) with the rectangular lower part, called the body (corpus sterni). Attached to the bottom of the body is a process made of cartilage called the xiphoid process that turns into bone (ossifies) as people age. The sternum is easily felt in the living, but often not remembered as a separate bone.

Generally, most vertebrae (singular: vertebra) are composed of two major segments, a body and a neural arch that attaches to the body at the pedicle (Figure 2.5B). The body is the major part of the bone, while the latter structure arches away from the body, forming the vertebral foramen that allows for passage of the nerve column down the spine. On most vertebrae, this arch has three processes; the spinous process extends posteriorly, while transverse processes extend laterally from the right and left sides. In addition to these structures, each vertebra has (at least) four areas, called facets, for joining (articulating) to other bones, two on top (superior) for joining with the bone above and two on the bottom (inferior) for the bone below. Lastly, the vertebrae of the middle rib cage contain extra facets for articulating with the ribs, both on the body (where they are called demifacets) and on the transverse processes.

There are three types of vertebrae, definable by their placement in the spine: cervical, thoracic, and lumbar (see Figure 2.6F). The cervical vertebrae comprise the top seven bones of the spinal column, corresponding to the neck. The uppermost of this set is called the atlas (Figure 2.6A), a uniquely constructed vertebra that has no body and appears ringlike when viewed from above or below. The superior surface of this bone has facets (superior articular facets) that contact the skull at the occipital condyles, while the facets of the inferior side (not shown) contact the second cervical vertebra, called the axis (Figure 2.6B). This latter bone also is distinct from other vertebrae in that it has a knob of bone called the dens or odontoid process that points upwardly from the body. The remaining cervical vertebrae are built along the general pattern of other vertebrae, in that they have bodies, neural arches, articular facets, and spinous and transverse processes. Figure 2.6C shows the bottom of one cervical vertebra; notice the tip of the spinous process is divided into two parts (bifurcated); the other vertebrae do not have this bifurcation. Additionally, the transverse

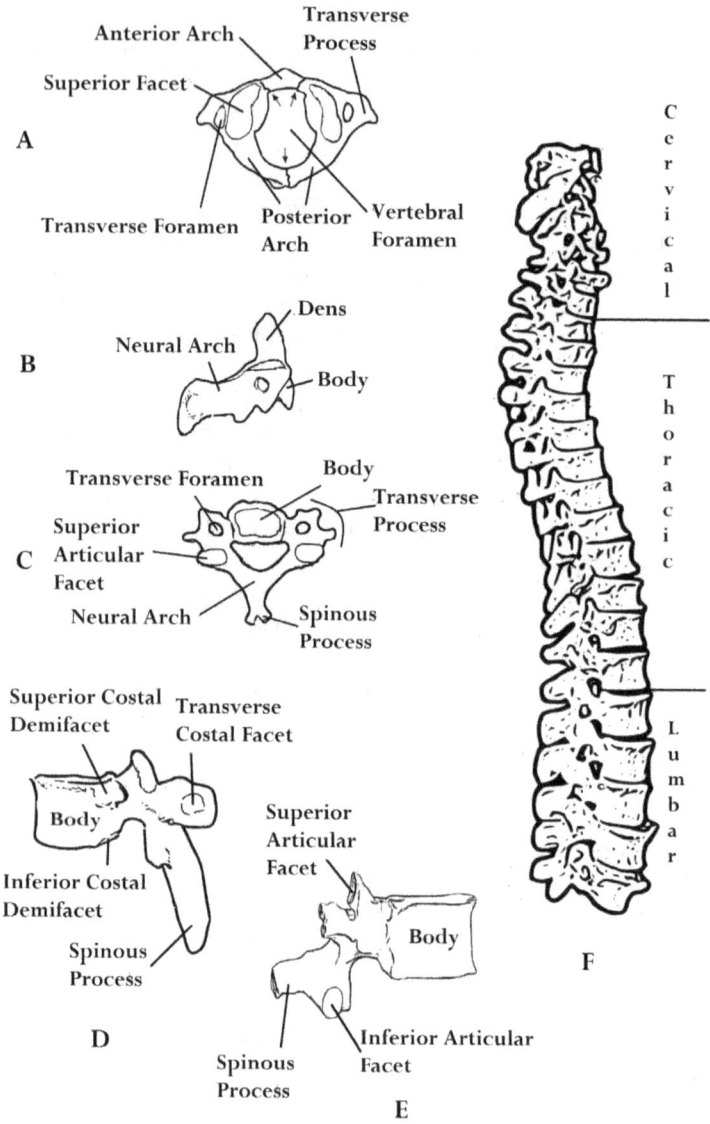

*Figure 2.6* Vertebrae and Vertebral Column: Atlas (A), Axis (B), Cervicle Vertebra (C), Thoracic Verebra (D), Lumbar Vertebra (E), and Assembled Vertebral Column Showing the Three Curves (F)

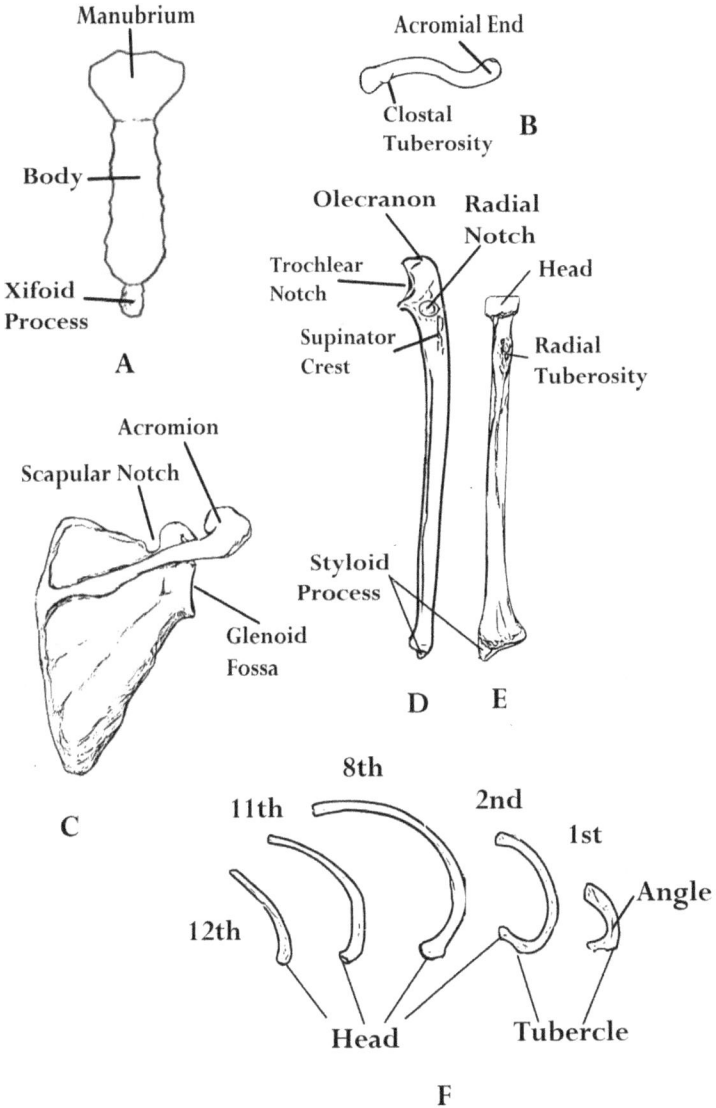

*Figure 2.7* Bones of the Axial and Appendicular Skeleton: Sternum (A), Clavicle (B), Scapula (C), Ulna (D), Radius (E), and 1st, 2nd, 8th, 11th and 12th Ribs (F)

process has a foramen (appropriately) called the transverse foramen, that is not found in other vertebrae.

The middle of the vertebral column is usually composed of 12 thoracic vertebrae that exhibit all the major structures described previously. These vertebrae have facets on both the body (called demifacets) and the transverse processes for articulation with the ribs (see Figure 2.6D). The long spinous processes of these vertebrae slant inferiorly to the point that they allow only limited posterior movement of the upper back. The lumbar vertebrae (Figure 2.6E) are the five bottom bones of the spinal column, above the sacrum. These do not have articular facets for the ribs and have shorter and wider transverse and spinous processes as well as mammillary processes that originate near the superior articular facets (not pictured). In addition, these have very large bodies.

The rib cage normally is composed of 12 ribs (singular: rib) on each side (men and women have equal numbers), all of which connect posteriorly to the vertebral column and most of which connect anteriorly to the sternum. Each rib has four basic components (Figure 2.7F), the head, neck, tubercle, and shaft. The head of the rib is the posterior part of the bone that has articular facets (actually two demifacets for all but the 1st, 10th, 11, and 12th ribs) for the body of the vertebrae, while further down the bone there is a facet that connects the ribs to the vertebral transverse processes. The neck is the section between the head and the tubercle. The shaft of the rib is the rest of the bone extending from the tubercle to the midline anteriorly, where most articulate with the sternum by way of cartilage (called the intercostal cartilage). The top two and bottom two ribs generally are easily distinguished. The first rib is small and flat (far right rib in Figure 2.7F); it articulates medially with the 7th cervical and first thoracic vertebrae and anterior-medially with the manubrium and sharply curves distal to the tubercle at the angle. The second rib is larger and less flat than the first but is still easily distinguished from others by its smaller size. The 11th and 12th ribs, also called floating ribs, can be identified by their short length and somewhat pointed ends that lack the squared-off appearance of other ribs (far left ribs in Figure 2.7F); these do not connect to the sternum.

APPENDICULAR SKELETON

The appendicular skeleton is composed of the bones of the upper and lower limbs. The upper limbs not only include the bones of the arms (called long limb bones) and hand but also those of the shoulder. Similarly, the lower limb bones include those of the legs (also called long limb bones) and feet but also those of the pelvis. The bones from each of these halves are described separately below.

*UPPER LIMBS*

Each shoulder girdle (right and left) is composed of two bones: the clavicle (collarbone) and the scapula. The clavicle (plural: clavicles), which is S-shaped when viewed from above or below (Figure 2.7B), articulates medially with the manubrium and laterally with the acromion of the scapula. Notice the costal tuberosity that points downward from the medial end. The triangular shoulder blade (Figures 2.7C), called the scapula (plural: scapulae), has three important features. First, there is the glenoid fossa (also called the glenoid cavity), which is the articular surface for the head of the upper arm bone (humerus). The second feature is the acromion, which has a facet for joining (articulating) to the lateral end of the clavicle. A final feature of this bone is the scapular notch that appears at the top (superior) end of the bone.

In addition to the clavicle and scapula, the upper limbs are composed of three long bones (called the humerus, ulna, and radius) and the bones of the hands. The humerus (plural: humeri) is the upper arm bone (Figure 2.8) that articulates proximally with the scapula (specifically the glenoid fossa) and distally with the ulna and radius. Its major features are the humeral head (for articulation with the glenoid fossa), the greater and lesser tubercles (knobs of bone near the head), the deltoid tuberosity (a mounded area on the lateral surface of the humerus near mid-shaft), medial and lateral epicondyles (boney mounds on the distal end), the capitulum and trochlea (a cylindrical articular area on the distal end), and the olecranon and coronoid fossas. The olecranon fossa accepts the olecranon from the ulna and prevents the elbow from bending backward (posteriorly) while the coronoid fossa accepts part of the ulna, allowing the elbow to be bent.

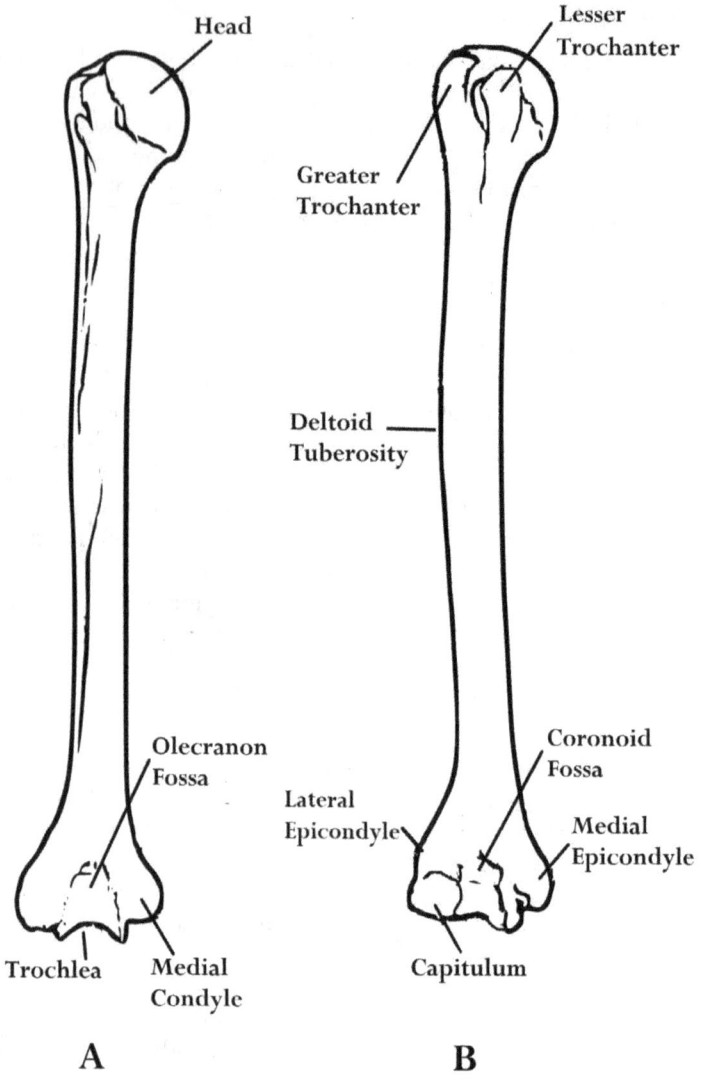

*Figure 2.8* The Humerus (Upper Arm Bone) in Back (A) and Front (B) Views

The ulna (plural: ulnae) is the medial bone of the lower arm (Figure 2.7D) when the skeleton is in the anatomical position. The trochlear notch articulates with the trochlea of the humerus proximally, and the ulnar styloid process of the distal end points toward the wrist. The radial notch is an oval shaped articular area just inferior to the trochlear notch that accepts the head of the radius. One last feature is the supinator crest which is where part of the supinator muscle attaches so that it can rotate the forearm and allow the hand to face forward.

The radius (plural: radii) is the outside bone of the lower arm (Figure 2.7E) when the body is in the anatomical position. The (cake pan-shaped) radial head articulates medially with the radial notch of the ulna and superiorly with the lateral portion of the trochlea of the humerus. The radial styloid process also points toward the wrist similar to that on the ulna. One last feature is the radial tuberosity that is for attachment of the biceps muscle.

The bones that make up the hand (Figure 2.11A) are those of the wrist (carpals), palm (metacarpals), and fingers (phalanges). There are 27 of these bones in each hand: 8 carpals, 5 metacarpals, and 14 phalanges.

*LOWER LIMBS*

The pelvis (Figure 2.1) is composed of four bones, a right and left os coxae (called 'hip bone' in Figure 2.1), the sacrum, and the coccyx. Each os coxae (plural: ossa coxae) in turn is composed of three bones: the ilium, ischium, and pubis (Figure 2.9A). These bones meet and join in the hip socket called the acetabulum (see dividing lines in acetabulum of Figure 2.9A) while the ischium and pubis join along the bottom of the bone. The acetabulum is where the head of the femur articulates with the os coxae, and the ischial tuberosity is the enlarged and roughened area of the ischium. When viewed medially (Figure 2.9B), the important features of os coxae are: the greater sciatic notch (a large notch formed by the posterior extension of the ilium and the superior part of the ischium), the symphyseal face (surface) of the pubic bone (the medial part of the pubic bone), a roughened L-shaped area, called the auricular surface on the inferior-posterior part of the ilium. This latter structure forms half of the sacroiliac joint since it is

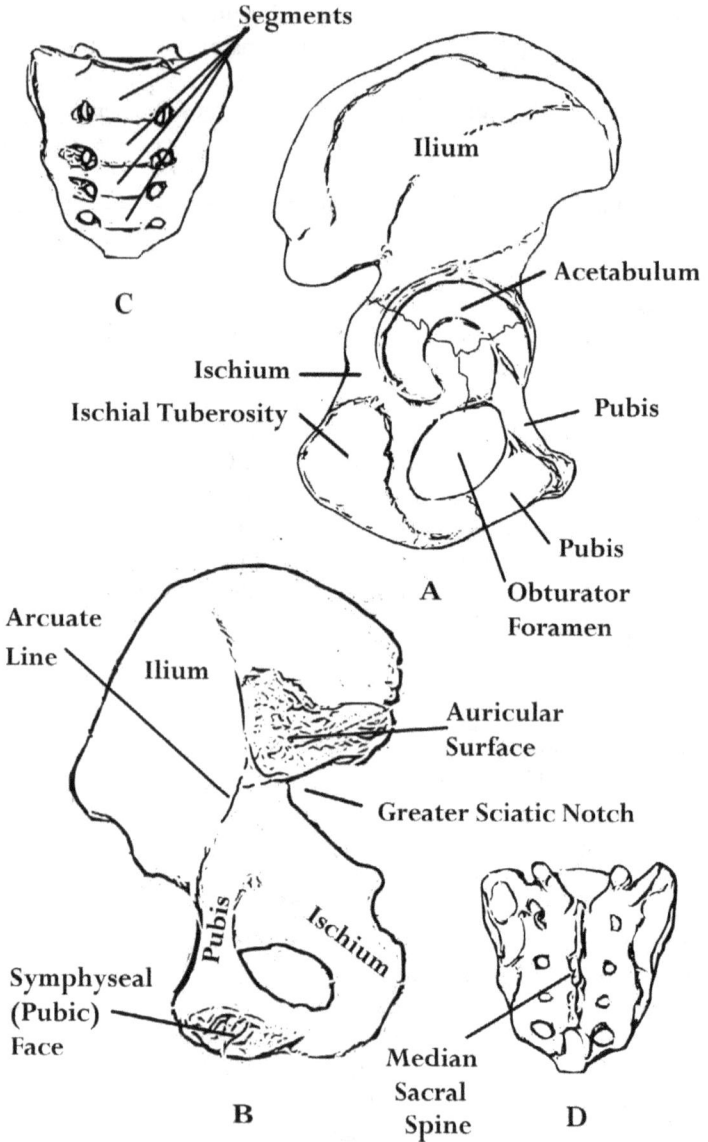

*Figure 2.9* Bones of the Pelvis: (A) Os Coxae Outside (Lateral) View; (B) Os Coxae Inside (Medial) View; (C) Sacrum, Front (Anterior) View; (D) Sacrum, Back (Posterior) View

where the sacrum attaches to the os coxae. Both the auricular surface and the symphyseal surface under-go changes through time, which makes them useful for the estimation of age at death. In addition to these structures is the preauricular sulcus. When present, this is a groove that runs along the superior margin of the greater sciatic notch on the ilium (not pictured). One last feature is the obturator foramen (Figure 2.9A) which is the opening formed by the ischium and pubis.

The sacrum (Figure 2.9C), which is the bone that attaches the vertebral column to the pelvis, usually is composed of five vertebrae that are fused both between the bodies (called segments) and between the transverse processes and the neural arches. The fused neural arches on the back (posterior) side of the bone form a ridge called the medial sacral spine (Figure2.9D). Finally, the coccyx (not pictured) usually is composed of four undeveloped vertebrae that fuse into a single bone later in life (i.e., after 25 or 30 years). This structure attaches to the inferior end of the sacrum.

Below the pelvis, each lower limb is composed of three long limb bones (the femur, tibia, and fibula), the kneecap (the patella), and the bones of the foot. Figure 2.10A displays the femur (plural: femora), or thigh bone, in the posterior view. Notice the ball-like end (the femoral head) for articulation with the hip socket (the acetabulum), the greater and lesser trochanters (knobs of bone), the linea aspera (roughened ridge along the posterior side), and the large medial and lateral condyles at the distal end (with a space between them called the intercondylar fossa) for articulation with the tibia. In addition, notice the medial and lateral supracondylar lines that converge into the linea aspera about a quarter of the way up from the distal articular end. Lastly, there is a roughened area of this bone distal to the greater trochanter for attachment of the gluteus maximus muscle (the gluteal tuberosity).

The tibia (plural: tibiae) has several features of interest (Figures 2.10C). First, the proximal end flares, forming the medial and lateral condyles for articulation with the distal femur; located between these is the intercondylar eminence, a raised area with two boney spikes. Also, the anterior surface displays the tibial tuberosity, a knob of bone that starts just below the condyles and merges with a boney ridge (the anterior crest) that extends inferiorly approximately two-thirds the length of the tibia. Finally, notice the process that extends inferiorly on the distal end; this is

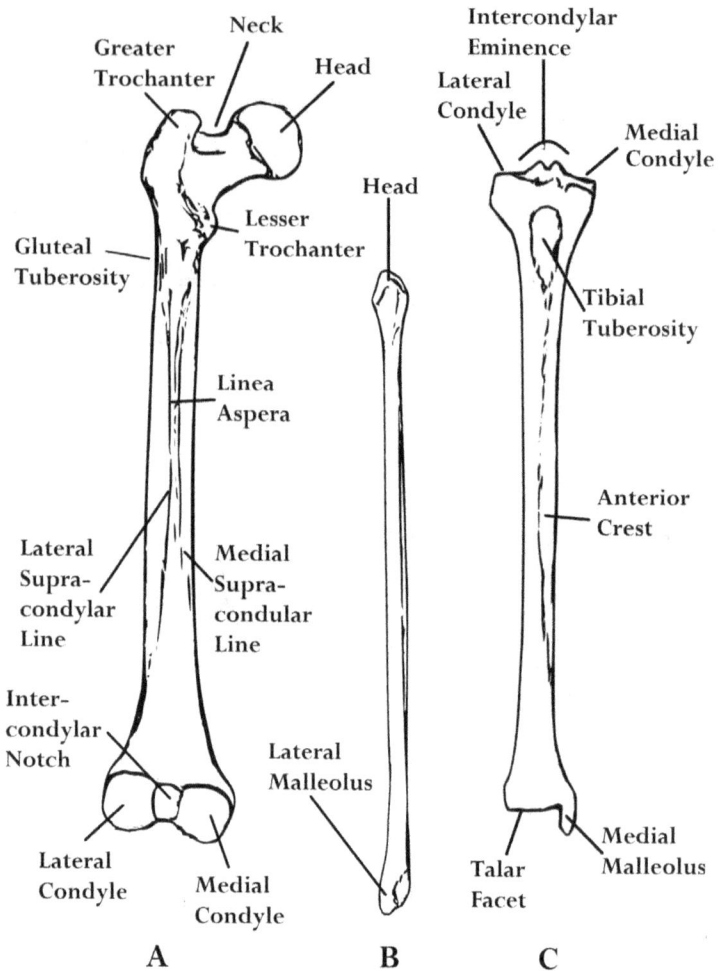

*Figure 2.10* The Long Limb Bones of the Lower Appendicular Skeleton: Femur, Rear (Posterior) View (A), Fibula (B), and Tibia, Front (Anterior) View (C)

the medial malleolus, which helps to stabilize the ankle by preventing sideways movement. On the bottom of the tibia is a facet for attaching to the talus, one of the bones that makes up the ankle.

The fibula (plural: fibulae) is a thin bone (Figure 2.10B) that articulates with the tibia on the lateral side. It is composed of a somewhat triangular head that contacts the tibia below the lateral condyle and a lateral malleolus to oppose the medial malleolus of the tibia. Finally, lying on the anterior aspect of the knee (covering the inferior femur and superior tibia) is the patella (plural: patellae), or kneecap (Figure 2.1).

The foot (Figure 2.11B) is composed of bones of the ankle (tarsals), foot (metatarsals), and toes (phalanges). The total number of bones in this structure is 26: 7 tarsals, 5 metatarsals, and 14 phalanges. Two of the tarsal bones are particularly important. The top bone is the talus, which has a superior articular surface (trochlea) and medial and lateral articular facets for malleolus's of the tibia and fibula. A feature of this bone is the posterior process, which is the most posteriorly projecting part of this bone. Just inferior to the talus is the calcaneus; it makes up the heel of the foot and articulates superiorly with the talus and anteriorly with many of the remaining bones of the ankle. It has two articular surfaces on the front that are usually connected: anterior articular facet and medial articular facet; occasionally the facets do not touch and are completely separate. Also, the lower back of this bone has a large tuberosity for attachment of Achilles tendon and other muscles.

One last group of bones that are seen in most, if not all, human skeletons are accessory ossicles called sesamoid bones. These small bones form in different muscles and tendons over various joints of the wrist, hand, and foot where they appear to reduce friction on the muscles and tendons. The largest of these is the patella, which is recognizable (although consisting of cartilage) by the $23^{rd}$ week of pregnancy and turns into bone (calcifies) around the $14^{th}$ to $16^{th}$ year of age. However, when fully developed, most sesamoid bones are anywhere from a few millimeters to over one centimeter in size. Since they are under (at least partial) genetic control, these ossicles can help show relationships between populations. Also, there is some evidence that persons who use their hands more than most other people have larger sesamoid bones.

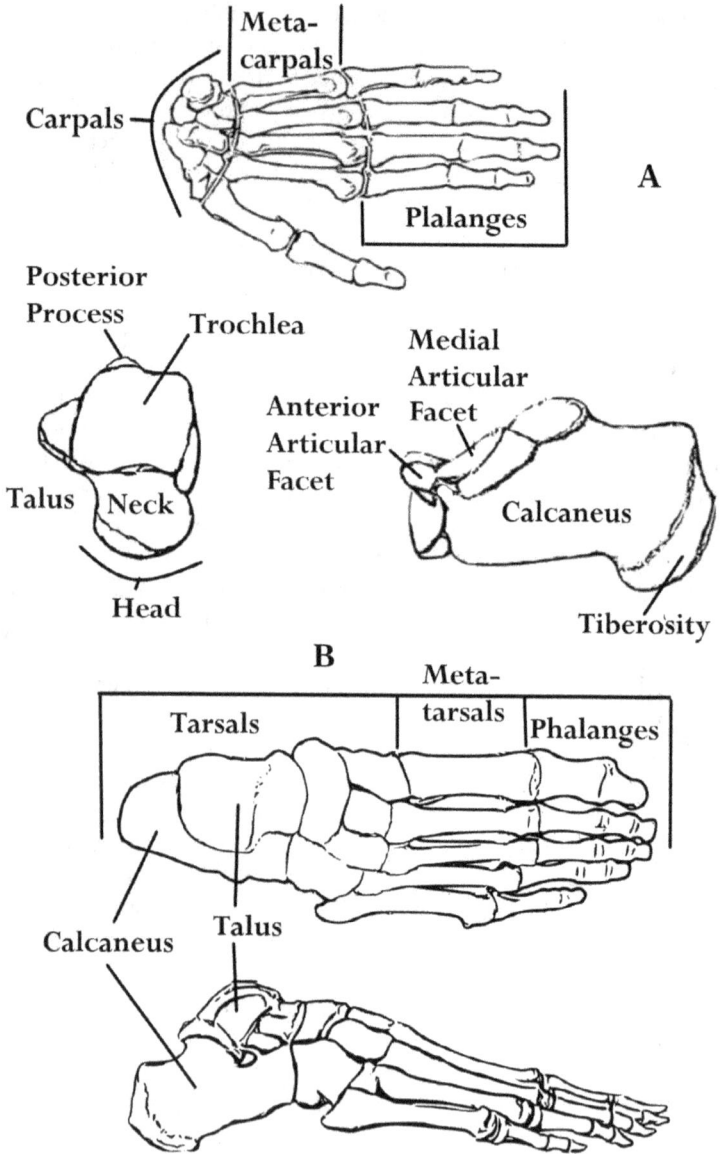

*Figure 2.11* Bones of the Hand and Foot: Hand (A), Foot in Two Views, Superior View (B, Top Image), and Side View (B, Bottom Image), Talus and Calcaneus Labelled on Figure

# BASIC TOPICS IN BONE BIOLOGY

Beyond recognizing the names of bones, skeletal biologists also understand basic bone anatomy and growth. Anatomy refers to aspects of bone beyond those described above; that is, the internal structures and even microscopic features of bones in general. In addition, the process by which bones grow also is important to know since the stages through which the skeleton passes give clues to the age of persons and the stresses they encountered during, and after, growth.

### ANATOMY OF BONE

The anatomy of bone can be thought of as having three areas: gross external, gross internal, and microscopic. Gross external anatomy are the components of a whole bone that are visible without magnification. These include the features described previously for individual bones, such as articular ends, prominences (tubercles and tuberosities), holes for blood vessels and other tissues (foramina), and openings (e.g., eye orbits, nasal aperture, external acoustic meatus). In addition, there are a number of lesser structures, such as grooves (indicating the presence of overlying blood vessels) and lines (for the attachment of muscles by tendons).

Besides these, three main gross structures of long bones are recognized (see Figure 2.12A): diaphysis, metaphysis, and epiphysis. The diaphysis (plural: diaphyses) is the shaft of the bone. It composes most of the total length and has a flared structure, called the metaphysis (plural: metaphyses), at each end. Each metaphysis is covered by an epiphysis (plural: epiphyses) that caps the end of the bone. This latter structure is separate from the other two during development, becoming fused after growth has finished. (This event, called epiphyseal union, is useful for estimating the age of individuals; see Chapter 5.)

The smooth exterior of all skeletal elements is made up of cortical (also called compact) bone, and the surface of this bone is referred to as the periosteal surface (so called because of a tissue called the periosteum that covers the bone and can lay down new bone). Cortical bone is composed of a strong, well-organized tissue called lamellar bone that is laid down in thin layers that run parallel

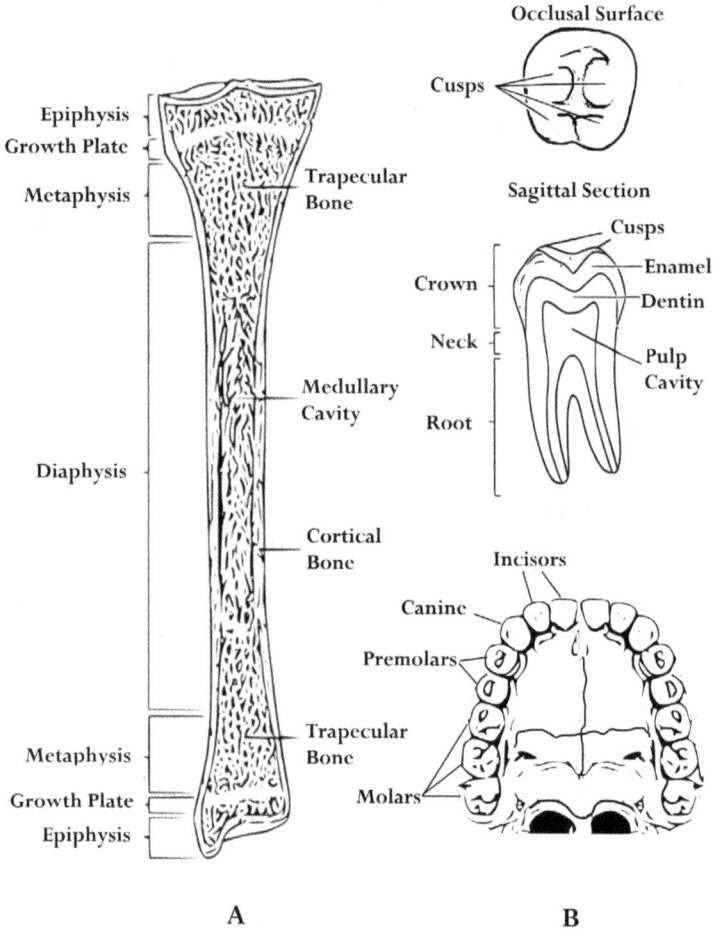

*Figure 2.12* Anatomy of Bone (A) and Teeth (B)

to the long axis of a bone. In contrast to the exterior cortical bone, or cortex, is the cancellous or trabecular bone of the interior. This spongelike structure occurs in the metaphyses of long bones, within the ribs and all bones of the hands and feet, inside the bodies of the vertebrae, and between the inner and outer cortical surfaces of the cranial vault (where it is called the diploe). Its primary function is to reinforce the bones without adding excess weight. A final internal structure is the medullary cavity or canal, which is the opening that runs through the center of all long bones. In life, it is filled with fatty tissues that are the target of carnivores when they gnaw on the skeleton after death. The interior surface of the medullary cavity is called the endosteal surface.

## OVERVIEW OF THE HUMAN DENTITION

Along with human bones, skeletal biologists also analyze human teeth. This analysis requires an understanding of dental anatomy, recognizing the types and placement of teeth, an understanding of how dentition develops, and an awareness of the variations that can occur within teeth. There are entire books written about these topics so only the most basic topics will be described here.

The basic anatomy of a tooth is illustrated in the top two diagrams in Figure 2.12B. The crown is the part that is visible above the gum; it is covered with white enamel that is peaked in some teeth into points called cusps. The place where the crown meets the rest of the tooth is termed the neck; this is the tapered area below the enamel that is slightly wider than the remaining tooth. The root is that part of the tooth that is embedded in the jaw; it is secured in place by a ligament that prevents it from falling out during chewing (mastication). The number of roots depends on the tooth; generally, the front teeth have only one, while those toward the back of the mouth have two, three, and even four of these structures. The root and that part of the crown covered by the enamel are both made up of a bonelike material called dentin.

When studying teeth, five basic directions (similar to the cardinal directions used in the body) are distinguished for the tooth surfaces. The mesial surface refers to those parts of the teeth closest to the midline, while distal refers to surfaces away from the midline. Also, the inner parts of teeth are referred to as lingual (toward the tongue), while the outer parts are either labial (near the lips) or buccal (near the cheeks). Finally, occlusal refers to the chewing surface of teeth.

Another convenience when studying teeth is to divide the mouth into quadrants: upper left and right, and lower left and right. Within each of these quadrants, the sequence of teeth is the same (see bottom drawing in Figure 2.12B); starting at the midline

(mesially) are the incisors, followed (distally) by the canine, premolars, and molars. The incisors are the flat, chisel-like teeth in the
front of the mouth that are easily visible when persons are smiling
or talking. Their crowns are wider (mesial-distally) than they are
thick (labial-lingually). Because they have only one root, they are
often lost post-mortem, which is unfortunate because these teeth
are valuable in estimating ancestry (see Chapter 3 "Estimating
Ancestry"). The canine is the pointed tooth next to the second
incisor. These single-cusped teeth, sometimes called "eye teeth" or
cuspids, also have a single root; however, it is so long that it
brackets either side of the nasal opening, anchoring it securely in
the jaws. The next teeth in line are the premolars (called bicuspids
by some dentists), which usually have two cusps and one or two
(usually joined) roots. The final, most distal teeth are the molars;
these are the square to rectangular chewing teeth found at the rear
of the mouth. The upper molars generally have three roots, while
the lower molars have two; these are often fused in both the upper
and lower third molars.

As most people know, humans have two sets of teeth during life,
the deciduous dentition (also called baby or milk teeth) and permanent teeth. Deciduous dentition is composed of five teeth per
quadrant: two incisors, one canine, and two molars (notice that
there are no deciduous premolars). This results in a total of $5 \times 4 =$
20 teeth. The permanent dentition is composed of eight per quadrant: two incisors, one canine, two premolars, and three molars; this
results in a total of $8 \times 4 = 32$ teeth. Occasionally, variations in
these numbers occur. Extra (supernumerary) teeth usually appear as
small, peg-like structures most often located in the area between the
normal teeth. Similarly, some teeth never erupt (e.g., the missing
wisdom tooth) or never develop (congenitally missing teeth).

Deciduous can be distinguished from permanent teeth by their
size and color. Since baby teeth are smaller than their adult coun-
terparts, it requires little experience to distinguish the two. Also,
the crowns of deciduous teeth often will have a slight yellowish
tint; this is caused by their thin enamel, which allows the color of
the underlying dentin to show through. However, the crowns of
permanent teeth generally are the white color of their enamel
because this substance is so thick that the underlying dentin is fully
hidden.

The sequence by which teeth form and erupt follows a relatively set pattern. Formation entails the coalescence of enamel into cusps, followed by apposition of this substance as well as dentin from the occlusal surface toward the tip of the root. When approximately one-half of the root is formed, the tooth emerges into the mouth through what is called the alveolar margin of the jaws, where it continues to grow until it reaches a genetically determined length. Thus, by knowing the schedule by which the different parts of the teeth form and the timetable for their eruption, the age of sub-adults can be estimated fairly accurately (see Chapter 5 "Estimating Age at Death").

## SUMMARY

1   The adult human skeleton normally is composed of 206 bones.
2   A number of cardinal directions and planes simplify the process of examining the human body.
3   Important features of the skull include the bones, the sutures that separate these bones, landmarks (used to identify various points and regions on the skull), and the sinuses found within various bones.
4   The axial skeleton is composed of the vertebrae (7 cervical, 12 thoracic, and 5 lumbar), the ribs (12 on each side), and the sternum.
5   The upper limbs are composed of the scapulae, clavicles, three long limb bones (humeri, ulnae, and radii) and the bones of the hand: wrist (carpals), palm (metacarpals), and fingers (phalanges).
6   The lower limbs are composed of the pelvis, three long limb bones (femora, tibiae, and fibulae), the kneecaps (patellae), and three bone sets at the end of the leg: ankle and heel (tarsals), foot (metatarsals), and toes (phalanges).
7   The pelvis is composed of four bones: right and left ossa coxae, the sacrum, and the coccyx; each os coxae likewise is composed of three bones that fuse in early adolescence: ilium, ischium, and pubis.
8   There are four types of teeth in the human mouth: incisors, canines, premolars, and molars.
9   Normal humans have two sets of teeth: deciduous (also called baby or milk teeth) and permanent.

## FURTHER READING

The best text on human skeletal biology is: TD White, MT Black, and PA Folkens, *Human Osteology*, 3rd ed. (Cambridge, MA: Academic Press, 2012). Others include WM Bass, *Human Osteology: A Laboratory and Field Manual*, 5th ed. (Springfield, MO: Missouri Archaeological Society, 2005), and DG Steele and CA Bramblett, *The Anatomy and Biology of the Human Skeleton* (College Station, TX: Texas A&M University, 1988).

# ESTIMATING ANCESTRY

When looking at people in the world, most persons recognize that differences can be seen and that these differences can be used to arrange people into different groups. In early times, skeletal (and other) researchers used the term 'race' to classify people into different groups but the linking of race with racism along with questionable theory and methods for estimating race led to this term being dropped. In an attempt to remove racism (and other negative emotions) from categorizing humans, terms like ethnicity and ethnic group as well as ancestry and ancestral group, and even population affinity have been used. For the purposes of this book, the fairly neutral term of 'ancestral group' is used to avoid the problems that come from 'race' and other terms.

When considering which ancestral groups to use in their work, most skeletal biologists use the terms of the US Census Bureau to categorize skeletal remains since these are known to most people (at least in the United States). Although it still uses the term race, the Census Bureau divides people into one (or more) of five major ancestral groups: Asian, American Indian or Alaska Native, Black or African American, Native Hawaiian or Other Pacific Islander, or White. Of these, the three main groups that can be most easily estimated from skeletal remains are: Asian, Black (African American), and White. Asian is the term used for those people who can trace their ancestry to eastern and southeastern Asia, especially China, Japan, Mongolia, and the Koreas; Native Americans are often considered part of this group as their ancestors came from eastern Asia. Black people can trace their ancestry to the people who live south of the Saharan desert, especially the counties of

DOI: 10.4324/9781003487944-3

west central Africa: Angola, the Congo, Gabon, Gambia, Sene-
gal, and Malie. Whites are people whose ancestry can be traced
to European countries like Great Britain, France, Italy, Germany, Nordic countries, and the countries of eastern Europe.
Alaska Natives and Native Hawaiians or Other Pacific Islanders
are more difficult to recognize from their bones and will not be
considered further.

This chapter discusses the visual traits and metric measurements
used to estimate the ancestral group of skeletons. Because the skeletons of some groups are smaller and more lightly built on the
average than those of other groups (e.g., southeast Asians versus
European Whites), the bones of males in 'small' groups may appear
to be females of 'large' groups. Since this could lead to difficulties
in estimating sex (see Chapter 4), estimating ancestry is the first
step in showing the biological profile of persons from their skeletal
remains. First, visual traits of the skull will be described, followed
by those of the postcranium. This will be followed by a description
of metric measurements useful for this purpose.

## VISUAL TRAITS

When estimating ancestral group from skeletal remains, the process is
similar to that described in Chapter 1 for estimating sex. That is, features of the bones are compared to those seen in a majority of the
individuals of one of the three major groups from skeletal collections
of known ancestral groups. Research has shown that the skull, especially the face, is the best osteological structure for estimating ancestry.
Early skeletal biologists noticed over a dozen traits of the skull that
differed between the ancestral groups and reported an accuracy rate of
85% to 90% in assigning ancestry to an unknown skull. Figure 3.1
illustrates some of these features in front and side views of the skulls of
Asians (including Native Americans), Blacks, and Whites.

Starting with the nose, notice that the widths of the nasal
opening (nasal aperture), when viewed from the front, vary from
wide (Black) to medium (Asian) to narrow (White). This is one of
the most useful characteristics for distinguishing between the three
ancestral groups. Also, from the front views (although difficult to
see in the Figure 3.1) is the shape of the nasal bones; in Blacks,
they are wide and rounded (sometimes described as looking like a

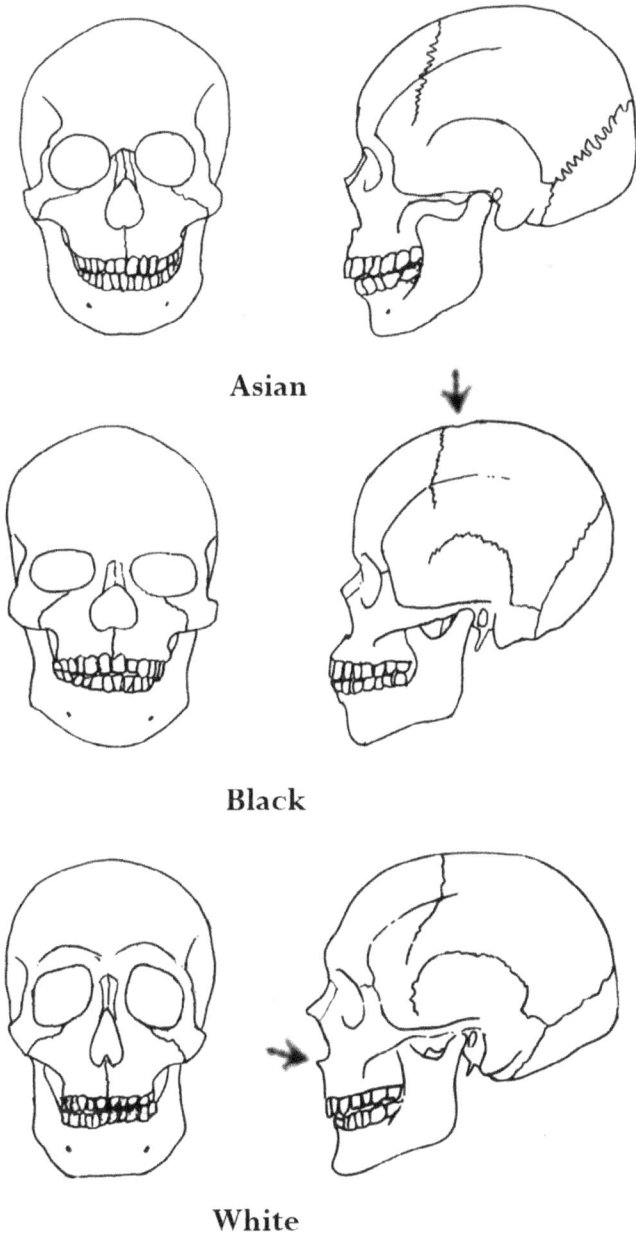

**Asian**

**Black**

**White**

*Figure 3.1* Front and Side Views of the Skulls of the Three Main Ancestral Groups

Quonset hut) while in Whites they are narrow and tented (as in a A-frame house) and are like a sagging tent in Asians. Next, from the side view, notice the sharp point of bone sticking out from the lower border of the nose in the White skull (arrow on side view of White skull in Figure 3.1). This is called the anterior nasal spine and is particularly large in Whites but smaller or absent in the other two groups. Also notice from the side view how much more the nasal bones jut out from the face of the White skull when compared to the other two groups. One last feature of the nose that deserves mentioning is the lower border of the nasal opening (piriform aperture). In Blacks, the floor of the nasal opening curves down into the front (anterior) of the upper jaw bones (maxillae) so that there is no obvious point where the floor ends, and the front of the maxillae begins. The term 'guttered' has been used to describe this contour. In Whites, there is a wall of bone (called the nasal sill) separating the piriform aperture and the front of the upper jaws that forms a definite break between the floor and maxillae. Asians are intermediate between the other two groups for this area of the face.

Moving on to the rest of the face, notice in the front views the different shapes of the eye orbits (see Figure 3.1). In Asians, they are fairly rounded but are rectangular in Blacks and angular (rhomboid) in Whites. Also, notice that the eye orbits in Blacks are more widely spaced than those of the other two groups (see parallel lines on facial views in Figure 3.2). The facial jutting seen in the side views of the three groups also differs. In Blacks, the jaws jut forward from the rest of the face, while the jaws are smaller and receding in Whites, and Asian jaws are somewhere between these two extremes (see dashed lines of side views in Figure 3.2). Also, the lower border of the eye orbits in Asians project in front of the upper border (see solid lines over the eye orbits of side views in Figure 3.2) while it is behind the upper border in Blacks and (especially) in Whites.

The braincase also differs between the three ancestral groups. Notice from the front and side groups that the browridges are larger in Whites than in the Asians and Blacks (see Figure 3.1). Also, the vault sutures, especially the lambdoid suture, are more complex in Asians than in the other two groups (not shown). In addition, there is often a dent behind bregma (post-bregmatic

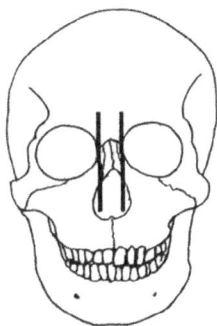
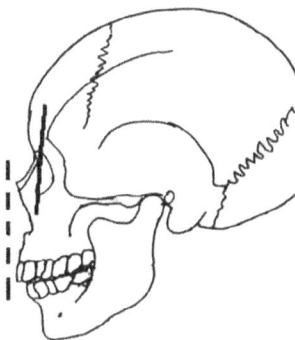

## Asian

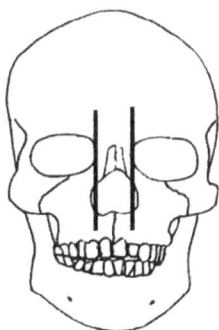
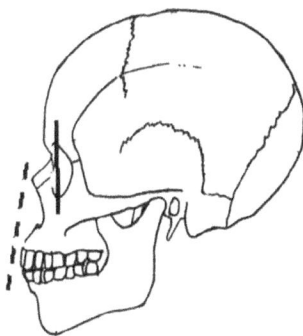

## Black

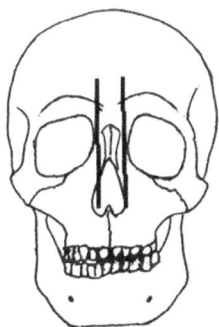
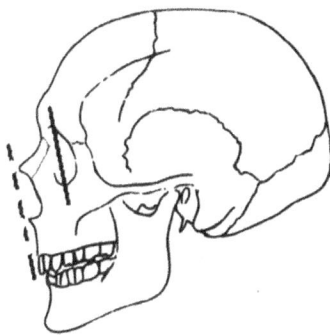

## White

*Figure 3.2* Front and Side Views of the Skull of the Three Main Ancestral Groups with Lines Emphasizing the Differences Seen Between the Groups for Various Traits (See Text for Description)

depression) in Blacks (see arrow on side view of Black skull of Figure 3.1) but is less common in the other two groups. One last trait is the shovel-shaped incisor. Seen particularly in the first upper incisors of Asian, the inner (lingual) surface has a ridge of enamel from the neck to the edge (occlusal surface) on either side (see left image in Figure 3.4A). These ridges are more rare in the other two groups where the lingual surface of the incisors are flat, like a spatula (see right image in Figure 3.4A).

The early osteologists mentioned above listed the traits that distinguished the three main ancestral groups without quantifying how often they appeared in one or more of the groups. As more research was done, the frequencies of each of these (and other) traits were computed which meant that the probability that a skull was from each of the groups could be given. Table 3.1 gives these figures for most of the traits described above. Looking at the frequencies/probabilities in the table, notice that none of the traits are found in one and only one of the groups; that is, none of the rows has a 1.00 in one of the columns but 0.00s in the other columns. This is because all of the trait categories are found in all of the groups with one group usually having a higher frequency/probability than the other two. For example, as stated above, the skulls of Whites have narrow nasal openings while those of Blacks are wide and Asians are in between these two extremes. Notice in Table 3.1 that the probability that a narrow nose is from a White skull is $p=.880$ (or 88%) but the remaining 12% of narrow noses is almost evenly divided between Blacks ($p=.062$) and Asians ($p=.059$). A similar, but less striking, situation occurs with wide noses, where the probability that it is from a Black skull $p=.678$ (approximately 68%) and the remaining amounts divided almost equally between Whites and Asians. These results show the 'messiness' of human skeletal data mentioned in Chapter 1 as well as the variability of the visual traits described above and the difficulty of estimating ancestral group from the skull.

As discussed in Chapter 1, a collection of frequencies/probabilities can yield overall probability of group membership (in this case, the estimation of ancestry) using all of the traits that can be seen in a skull. Table 3.1 lists six traits with two to five expressions that could be combined to give a rough estimate of the probability that a skull is Asian, Black, or White. For example, suppose the

*Table 3.1* Frequencies of Visual Characteristics of the Skull of the Three Main Ancestral Groups in the United States[1]

| Trait | Category | Frequency/Probability | | |
|---|---|---|---|---|
| | | Asian | Black | White |
| Anterior Nasal Spine | Small | 0.549 | 0.300 | 0.150 |
| | Medium | 0.150 | 0.423 | 0.427 |
| | Large | 0.102 | 0.287 | 0.612 |
| Inferior Nasal Aperture | Strong Slope | 0.359 | 0.602 | 0.039 |
| | Slight Slope | 0.345 | 0.565 | 0.090 |
| | Right Angle | 0.530 | 0.258 | 0.212 |
| | Small Sill | 0.063 | 0.229 | 0.707 |
| | Large Sill | 0.083 | 0.183 | 0.733 |
| Interorbital Breadth | Narrow | 0.452 | 0.139 | 0.409 |
| | Medium | 0.418 | 0.236 | 0.345 |
| | Broad | 0.095 | 0.742 | 0.163 |
| Nasal Aperture Width | Narrow | 0.059 | 0.062 | 0.880 |
| | Medium | 0.485 | 0.258 | 0.257 |
| | Broad | 0.142 | 0.678 | 0.180 |
| Nasal Bone Contour | Low, Rounded | 0.424 | 0.505 | 0.070 |
| | Oval | 0.395 | 0.354 | 0.251 |
| | Steep Walls, Flat Superior Surface | 0.692 | 0.143 | 0.165 |
| | Steep Walls, Narrow Superior Surface | 0.173 | 0.269 | 0.559 |

| Trait | Category | Frequency/Probability | | |
|---|---|---|---|---|
| | Triangular Cross-Section | 0.023 | 0.398 | 0.579 |
| Post-Bregmatic Depression | None | 0.422 | 0.239 | 0.340 |
| | Present | 0.113 | 0.569 | 0.318 |
| Shovel-shaped Incisors | Absent | 0.091 | 0.446 | 0.462 |
| | Present | 0.811 | 0.111 | 0.077 |

[1] Skull traits combined from information in Hefner (2009) and Klales and Kenyhercz (2015); shovel-shaped incisor data combined from Hinkes (1990). See Further Readings for full citation.

bones around the nose of a specimen are found that have the characteristics of a White skull described by early researchers. Thus, from the frequencies/probabilities in Table 3.1, it has a large nasal spine (.l02,.287,.612), a large nasal sill (.083,.183,.733), narrow nasal opening (.059,.062,.880), and A-frame nasal bones/triangular cross section (023,. .398,.579). The totals for each of these traits from Table 3.1 are: Asian=(.102+.083+.059+.023)=.267, Black= (.287+.183+.062+.389)=.921, and White=(.612+.733+.880 +.579)=2.804. A rough estimate of probability that the nose bones belong to each of the three groups is: Asian=.267÷4=.067, Black=.921÷4=.230, and White=2.804÷4=.701, and since the White probability is greater than the other two groups, the nose bones are most probably from a White person.

Although the above discussion and the frequencies/probabilities in Table 3.1 present the most useful visual characteristics for distinguishing the three main ancestral groups, a number of other traits also can help with this estimate (for a complete set of these, see Rhine, 1990 in the Further Readings section). For example, Asians have a more vertically oriented ascending ramus of the mandible than the other two groups. Also, extensions of the enamel beyond the molar crown onto the neck (called enamel pearls) are more common in Asians than in Whites and Blacks.

Although the bones of the postcranium are not very useful for estimating ancestry, the shaft (diaphysis) of the femur is an exception in that it shows three different contours. Blacks generally have straight femoral shafts (Figure 3.3A) while those of Asians curve forward (anteriorly) as in Figure 3.3C with Whites being in between these two extremes (Figure 3.3B). Although this is only a partial list, knowledge that miscellaneous characteristics do exist can help skeletal biologists when analyzing osteological remains for ancestral group membership.

## METRIC TRAITS

Along with visual assessment, ancestry can also be estimated by taking various measurements of the skull and other bones. Metric methods have been studied since the early 1900s (in pre-computer times) as a means to estimate ancestry, but most of these methods are no longer used. Currently, the most useful metrics for this purpose are direct measurement, indexes, and discriminant functions. These will be described in the following section. However, one thing needs to be mentioned. Metric methods will result in an estimate of the most probable group for a given measurement or set of measurements. Unfortunately, this leads to a belief that these methods are more accurate than the visual methods and their frequencies/probabilities described above. This is not the case since the metric methods themselves have percent correct group assignment and percentage of misassignments. This should be kept in mind while reading the rest of this chapter.

One of the few direct measurements that has been shown to be useful for estimating ancestry is the angle of the intercondylar shelf of the distal femur. This feature is formed by the angle between the roof of the intercondylar notch (fossa) and the long axis of that bone (see Figure 3.3D). When x-rayed, the shelf appears as a bright (radiopaque) line on radiographs and its angle with the femoral axis is measured. Whites have a higher angle than Blacks, with the sectioning point between the two being 141°; values higher than 141° are more likely White, while angles below that are more likely Black. Since only 18% of the sample in studies overlap for this feature, there is an over 80% chance that ancestral group will be correctly estimated from this trait. However, since

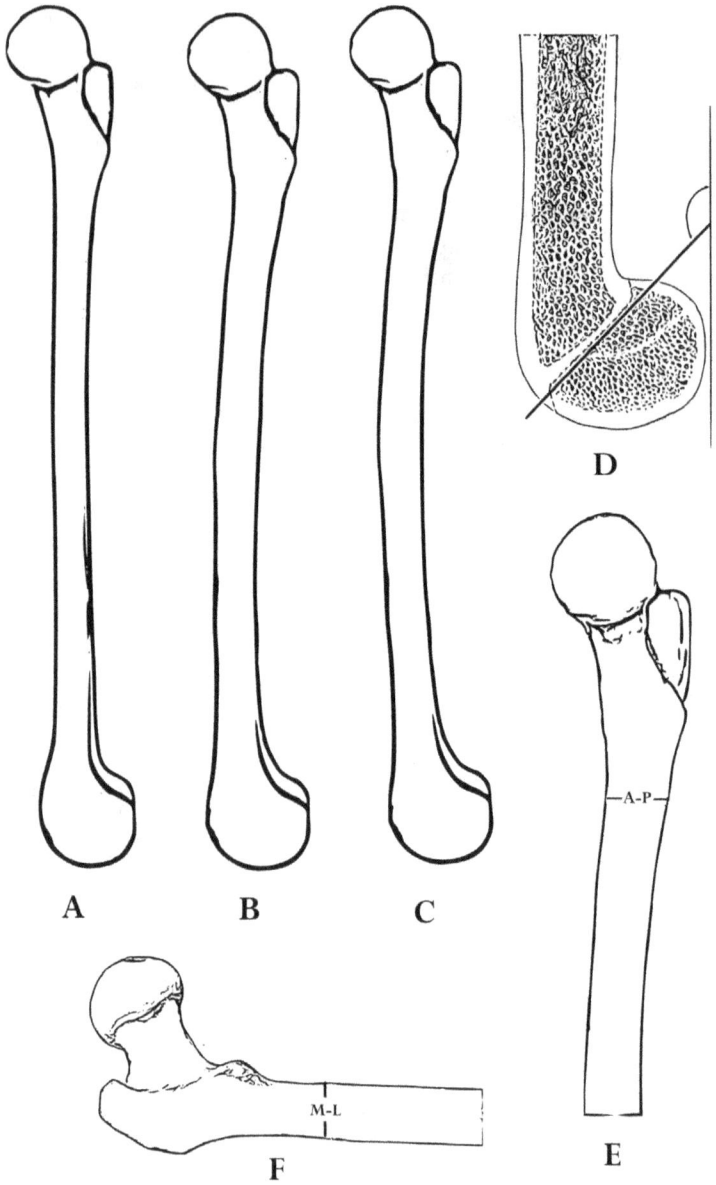

*Figure 3.3* Traits of the Postcranium that are Useful for Estimating Ancestral Groups (See Text for Explanation)

the frequencies of this feature are known only for Blacks and Whites, this method can be used only when there is good evidence that the skeleton is from one of these ancestral groups.

Several studies have shown that the shape of the femur just below (distal to) the lesser trochanter can help with estimating ancestry. These studies found that Native American femurs are flatter front-to-back (that is, oval shaped) than Blacks and Whites whose femurs are more rounded just below the lesser trochanter. This observation has been quantified using the front-to-back (anterior–posterior or A–P) and side-to-side (medial–lateral or M–L) measurements from this area (see E and F in Figure 3.3). These measurements have been combined into the Platymeric index (PI), which is: PI = (A-P ÷ M-L) × 100. This index has a sectioning point just like discriminant functions described in Chapter 1. In this case, PIs over 84.3 are more likely to come from the femurs of Whites and Blacks while values below that sectioning point are more likely Native American. Unfortunately, this index cannot be used to separate White femurs from those of Blacks.

The last, and most commonly used, metric method is discriminant function analysis (DFA). As described in Chapter 1, this statistical procedure uses any number of measurements to distinguish two or more groups. Because of the large number of computations, DFAs only became easy to calculate when computers and software packages that included DFA programs became more available to researchers. The earliest functions (from 1962) used eight skull height, width, and length measurements (see Figure 3.4B) to compute two formulas: one that could be used to distinguish the male skulls of Whites, Blacks, and Asians (Native Americans) and the other for female skulls (they provided a discriminant function to estimate sex if sex was unknown). The measurement values were entered into the appropriate DFA sex equation, and the value computed by hand. This value was compared to two sectioning points, one between Whites and Blacks, and one between Whites and Native Americans to arrive at an estimation of group membership. These formulas reported being correct from a low of 75% of the time to a high of over 90% of the time. Since then, computer programs have been developed using modern skeletal samples that only require a researcher to enter skull values and group membership is calculated.

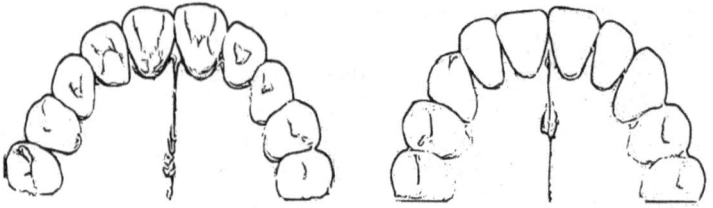

**A**

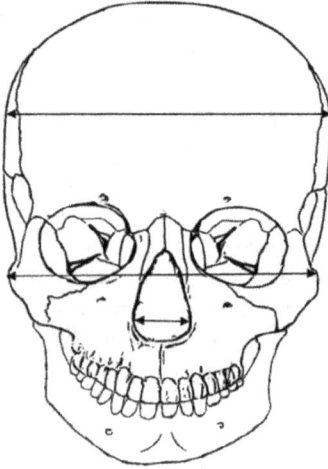

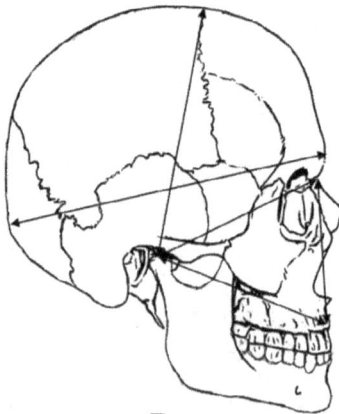

**B**

*Figure 3.4* Shovel-shaped Incisors (A) and Metric Measurements (B) Used to Estimate Ancestral Group (See Text for Explanation)

# SUMMARY

1   Since 'race' is associated with racism and has questionable scientific validity, the term 'ancestry' is used to place skeletons into one of the three main groups known to most people: Asian (including Native American), Black, and White.

2   The skull has most of the visual traits, such as nose and face form, that are useful for estimating ancestral group.

3   The femur is one of the few postcranial bones that shows differences between the ancestral groups.

4   There is one measurement (angle of the intercondylar shelf) and one index (Platymeric Index) of the femur that can be used to estimate ancestry.

5   Discriminant Function Analysis (DFA) has been, and is being, used to estimate ancestry from skull measurements.

# FURTHER READING

For general discussions of ancestry estimation, see relevant chapters: S Byers and C Juarez, *Introduction to Forensic Anthropology* (New York: Routledge, 2023), A Christensen, N Passalacqua, and E Bartelink, *Forensic Anthropology: Current Methods and Practice*, 2nd edition (London: Academic Press, 2019). For articles on estimating ancestral groups: G Gill and S Rhine, eds. *Skeletal Attribution of Race* (Albuquerque, NM: Maxwell Museum of Anthropology, Anthropological Papers No. 4, 1990). For list of skull ancestral group traits: S Rhine, Non-Metric Skull Racing. In: G Gill and S Rhine, eds. *Skeletal Attribution of Race* (Albuquerque, NM: Maxwell Museum of Anthropology, Anthropological Papers No. 4, 1990). For data in Table 3.1: JT Hefner, Cranial non-metric variation and estimating ancestry. (*Journal of Forensic Sciences*, 54:985–995, 2009); AR Klales and MW Kenyhercz, Morphological assessment of ancestry using cranial macromorphoscopics (*Journal of Forensic Sciences*, 60:13–20, 2015); MJ Hinkes, Shovel shaped incisors in human identification. In: G Gill and S Rhine, eds. *Skeletal Attribution of Race* (Albuquerque, NM: Maxwell Museum of Anthropology, Anthropological Papers No. 4, 1990). For ancestry from femur shaft: TD Stewart, Anterior femoral curvature: Its utility for race identification (*Human Biology*, 34:49–62, 1962). For ancestry from angle of intercondylar shelf: EA Craig, Intercondylar shelf angle: A new method to determine race from the distal femur (*Journal of Forensic Sciences*, 40:777–782, 1995). For ancestry from Platymeric Index: DJ Wescott, Population variation in femur subtrochanteric shape (*Journal of Forensic Sciences*, 50 (2):286–293, 2005). For early discriminant functions for ancestry from skull measurements: E Giles and O Elliot, Race identification from cranial measurements (*Journal of Forensic Sciences*, 7:147–157, 1962).

# ESTIMATING SEX

The second part of the biological profile is assessing the sex of persons represented by skeletal remains. There are a large number of both visual traits and measurements of the skeleton that help to estimate this characteristic. Early work on human skeletal analyses listed no less than 13 visual traits of the pelvis that help separate the sexes, but over time research has shown that three of them are most useful for this purpose. The same situation occurred with the skull; of the approximately 14 traits that distinguished male from female skulls, 4 have been shown to be most useful. Metric measurements also have been used to distinguish the sexes, especially measurements of the skull (e. g., length, width, height), with ranges of values for males versus females (e.g., in a sample from South Africa, skulls of length 185 mm or more are likely male, while skulls 178 mm or less are likely female). Later, these same measurements were entered into discriminant functions described in Chapter 1 to calculate an estimate of sex.

This chapter discusses the visual traits and metric measurements used to distinguish female skeletons from those of males. Most of these are useful only for adults; that is: persons who were approximately 18 years of age or older when they died. First, visual traits of the pelvis will be described, followed by those of the skull. Next, metric measurements useful for this purpose will be presented, followed by a discussion of sexing subadults (e.g., infants, children, adolescents).

## VISUAL TRAITS

The pelvis is the skeletal structure with the most information on the sex of the living adult individual. This is because the female

DOI: 10.4324/9781003487944-4

pelvis is designed for childbirth, something that is not needed by males. As noted above, a number of visually identifiable traits that differ between men and women have been identified. As illustrated in the images of Figure 4.1A, the male pelvis is tall and narrow as well as rugged (due to larger muscle markings) while in females it is comparatively shorter and broader as well as small and gracile. When a pelvis is viewed from above (superiorly), males can be distinguished from females by the shape of the pelvic inlet. This is formed by the top (superior) edge of both pubic bones, the arcuate lines of the ilia (see Figure 2.9B), and the top (superior) front (anterior) edge of the sacrum. In males, the pelvic inlet is heart-shaped, and therefore more restricted than the female. This opening in females is elliptical and therefore more spacious for newborns to travel through during birth. In Figure 4.1A, these two shapes are emphasized by the inserted drawings of a heart and an ellipse. The subpubic angle, formed by the lower (inferior) borders of the pubis and ischium, also differs between men and women. This feature is narrow and V-shaped in males but wide and U-shaped in females. The dashed lines in Figure 4.1A emphasize these shapes. This also is related to childbirth; the U-shape of females allows for easier passage of the infant during birth.

There are several other traits of the pelvis that distinguish the sexes. The pubic bone in males is more rectangular (stippled rectangle in Figure 4.1A, left side) while more squared in females (stippled square in Figure 4.1A, right side). Also, the obturator foramen in males is large and oval while it is small and triangular in females (not pictured). The greater sciatic notch, pictured in Figure 4.1B, shows the difference between males and females, with males having narrow notches (dashed line) while it is wide in females. The wider nature of this structure in females helps to separate the ossa coxae, again giving more room for childbirth. The upper edge of the greater sciatic notch (arrow on right image of Figure 4.1B) also helps separate the sexes. In females, there is a groove that runs along the bottom (inferior) edge of this upper (superior) border that is found in females but not found as often in males. Lastly, the sacrum of males is long and narrow while it is short and broad in females (not pictured); this too appears to be related to childbirth as the wider sacrum of females expands the distance between the two ossa coxae to provide more room for the passage of infants.

Male                                    Female

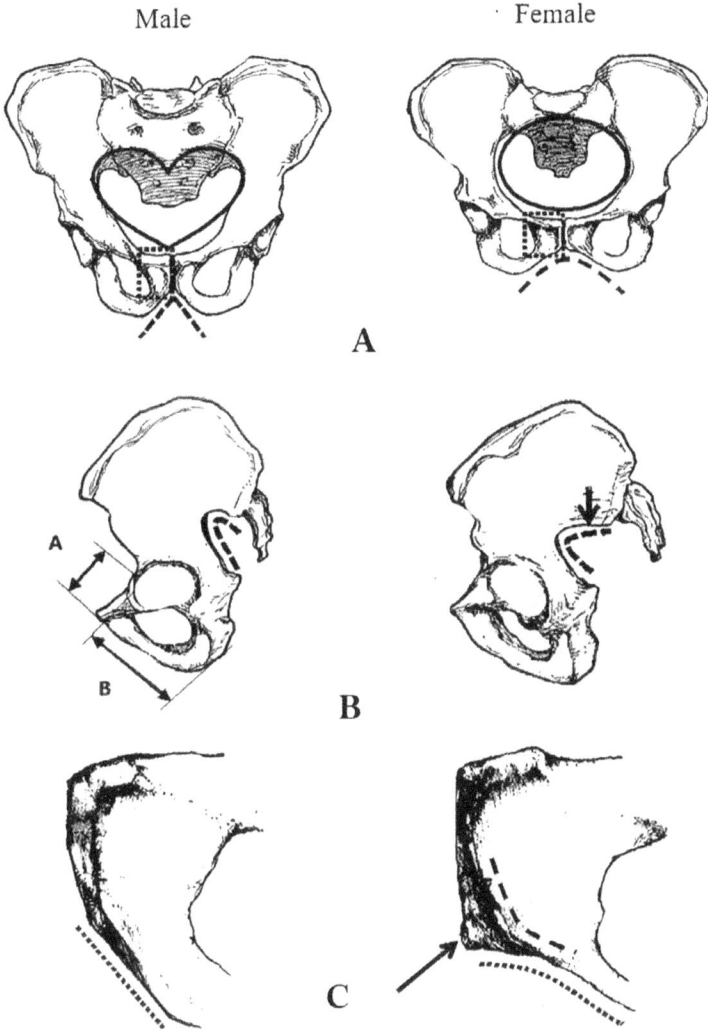

A

B

C

*Figure 4.1* Features of the Pelvis that Aid in Estimating Sex (See Text for Description)

For many years the above discussed pelvic features (and others) were used to differentiate male from female skeletons. However, research showed that three features of the pubic bone proved to be more accurate and easier to use than those described above. The bottom images of Figure 4.1 illustrate two of these three features. The first is the ventral arc. In females, this is a ridge that curves across the front (anterior) face of the pubic bone leaving a triangular section of flat bone in the lower center (inferior-medial) corner. On the right side of Figure 4.1C, the dashed line delineates this arc in females while an arrow points to the triangular section of flat bone. In males (left image of Figure 4.1C), the ventral arc is either absent or is small and closely hugs the center (medial) edge of the pubis, thereby not leaving room for a triangular area. The second feature is the subpubic contour which is a simplification of the subpubic angle discussed above. Rather than needing both pubic bones, research showed that the bottom edge of the ischium and pubic bones differ between males and females where it is curved upward (superiorly) in females while it is straight in males. The dotted lines in Figure 4.1C highlight this difference. The last feature is the central (medial) edge of the ischiopubic ramus (not pictured). In females, this can be narrow making it somewhat sharp (particularly on the superior part), while in males it is more rounded and blunter.

As more studies were done, researchers realized that these three traits are not simply present or absent, curved or straight, sharp or blunt. This led to the development of a 5-point scale for varying degrees of each of them. Table 4.1 lists the scales for the three traits as well as the probability of each being female or male. Similar to the situation of supra-orbital ridges discussed in Chapter 1, the frequency/probability that, say, a female has the most extreme example of a ventral arc (as shown in Figure 4.1) is very high ($p=.989$) while males rarely have this type of arc ($p=.011$). Conversely, at the other end of the scale, males are more likely to have no ventral arc ($p=.985$) while it is rarely missing in females ($p=.015$). The same situation can be seen in the subpubic contour and ischiopubic ramus; the extreme examples of each trait (i.e., category 1 and category 5) are most likely female and male, respectively. One finding of studies of these three traits is that the subpubic contour in males is not just straight but can be convex (i.e., curved downward). It is unknown why earlier studies did not note this.

*Table 4.1* 5-Point Scales for the Three Pubic Bone Features[1]

| Category | Description | Male | Female |
| --- | --- | --- | --- |
| Ventral Arc | | | |
| 1 | Well Defined | .011 | .989 |
| 2 | Defined | .084 | .916 |
| 3 | Small | .716 | .284 |
| 4 | Barely Visible | .968 | .032 |
| 5 | Absent | .985 | .015 |
| Subpubic Contour | | | |
| 1 | Deeply Concave | 0.007 | 0.993 |
| 2 | Slightly Concave | 0.082 | 0.918 |
| 3 | Straight | 0.717 | 0.283 |
| 4 | Slightly Convex | 0.949 | 0.051 |
| 5 | Convex | 0.984 | 0.016 |
| Medial Ischopubic Ramus | | | |
| 1 | Sharp | 0.016 | 0.984 |
| 2 | Dull | 0.079 | 0.921 |
| 3 | Rounded | 0.502 | 0.498 |
| 4 | Wide | 0.858 | 0.142 |
| 5 | Very Wide | 0.990 | 0.010 |

[1] Modified from Table 8.6 in Byers and Juarez (2023). See Further Reading for full citation.

The skull also displays features that help distinguish males from females. As described in Chapter 1, the simplest feature is overall size: males are larger than females, on the average. In addition, the larger skulls of males are more rugged looking with stronger muscle markings. Figure 4.2 illustrates more differences between sexes. The top images show the supra-orbital ridges discussed in Chapter 1. These boney mounds, which arch over each eye orbit, are also called supra-orbital tori, or brow ridges. They are small, or totally lacking, in females while they are large in males. The next trait down shows the ruggedness of male skulls as seen in the occipital where a ridge of bone, called the nuchal crest described in Chapter 2, can appear in males but not often in females. Also, as described in Chapter 2, a bump in the midline can develop into a

Female                                          Male

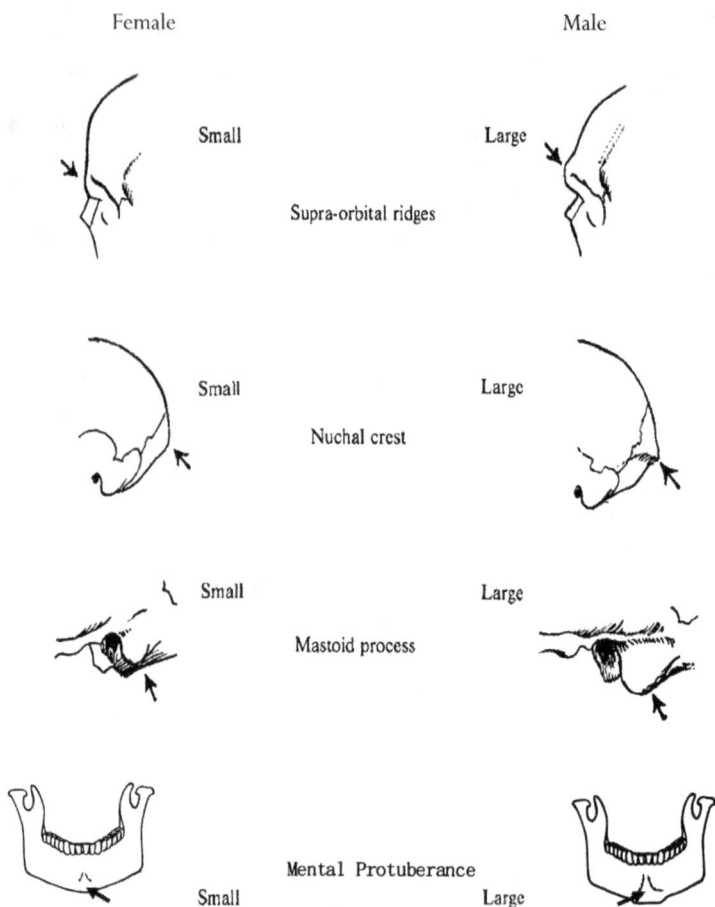

Small                          Large

Supra-orbital ridges

Small                          Large

Nuchal crest

Small                          Large

Mastoid process

Mental Protuberance
Small                                Large

*Figure 4.2* Features of the Skull that Aid in Estimating Sex. See Text for Description. Image from Figures 3.1 of *Digging Up Bones* by DR Brothwell (1981) Ithaca, NY: Cornell University Press (Used With Permission).

finger of bone projecting downward (inferiorly), called the external occipital protuberance or inion hook (because it appears at inion). The next trait shows that males have larger and more projecting mastoid processes than females, a trait that is often considered the most useful feature for distinguishing the sexes. The last structure is the mental eminence (also called the boney chin). In females, this is small and does not project forward very far on the mandible, while it is large and projecting in males.

As with the three traits of the pubic bone, research showed that the four traits of the skull also had more than just two expressions (i.e., large vs. small). Table 4.2 shows the 5-point scales developed for these

*Table 4.2* Cranial Trait Frequencies[1]

| Category | Description | Males | Females |
| --- | --- | --- | --- |
| Supra-Orbital Ridges | | | |
| 1 | Very Small | 0.149 | 0.851 |
| 2 | Small | 0.388 | 0.612 |
| 3 | Medium | 0.700 | 0.300 |
| 4 | Large | 0.826 | 0.174 |
| 5 | Very Large | 0.957 | 0.043 |
| Nuchal Crest | | | |
| 1 | Very Small | 0.247 | 0.753 |
| 2 | Small | 0.382 | 0.618 |
| 3 | Medium | 0.641 | 0.359 |
| 4 | Large | 0.763 | 0.237 |
| 5 | Very Large | 0.700 | 0.300 |
| Mastoid Process | | | |
| 1 | Very Small | 0.211 | 0.789 |
| 2 | Small | 0.372 | 0.628 |
| 3 | Medium | 0.684 | 0.316 |
| 4 | Large | 0.880 | 0.120 |
| 5 | Very Large | 1.000 | 0.000 |
| Mental Eminence | | | |
| 1 | Very Small | 0.000 | 1.000 |
| 2 | Small | 0.342 | 0.658 |
| 3 | Medium | 0.628 | 0.372 |
| 4 | Large | 0.776 | 0.224 |
| 5 | Very Large | 1.000 | 0.000 |

[1] Modified from Table 8.8 in Byers and Juarez (2023). See Further Reading for the full citation.

traits as well as the frequencies of each in males and females. Notice that some of the extreme expressions of the traits are found in one, and only one, sex (e.g., very large mastoid processes and very large mental eminence are found only in males). This is a rare finding that is probably due to sampling error; more research would probably reveal probabilities less than p=1, but still quite large.

After the pelvis and the skull, the remaining bones of the skeleton help to distinguish the sexes mainly by size and rugosity. Large males with rugged muscle markings can be easily distinguished from small skeletons with smooth bones. These criteria are most useful when the general size of a population is known. According to worldwide data, males in the Netherlands average around six feet and females around five and a half feet, while in Laos males are around five foot three inches and females are around five feet. Thus, a male skeleton from Laos is more the size of a female in the Netherlands than a male. Similarly, a female skeleton from the Netherlands is more the size of a male skeleton from Laos than a female from Laos. This illustrates the importance of knowing the size of the population of origin of skeletons (i.e., the ancestral group as discussed in Chapter 3). When population of origin is known, it is easier to use size differences to distinguish males from females.

## METRIC MEASUREMENTS

In addition to the visual traits just described, sex can be estimated from metric measurements taken on the skull, pelvis, and other bones of the skeleton. The most obvious way is to use these measurements is to assign the largest bones to males and smallest to females. As noted above, populations vary in height, and this also means they vary in all bone measurements. Thus, by measuring various bones from skeletons of known population and sex, the average (statistically called the mean) measurements of males and females as well as their variation can be calculated, thereby providing standards to use with skeletons of unknown sex. Using the example from Chapter 1, the average (mean) breadth of the epiphysis of the proximal tibia of Euro-American (White) males is around 79 mm while it is approximately 69 mm for Euro-American females. This and information on the variation of this measurement in each of the sexes led to the statistic of 74 mm as the dividing point between females and males. As described in Chapter 1, female sex is assigned to tibias under 74 mm for this measurement and male for tibias over that amount, resulting in correctly sexing 90% (p=.90) of tibias from skeletons of unknown sex. There are many other skeletal measurements that can be similarly used to estimate sex, although they are less accurate than the 90% given above. Many studies have been performed over the last decades using lengths, widths, heights, angles, and weights to estimate the sex of persons from their skeletal remains.

Another statistic that has been used to estimate sex is the index. This is calculated by dividing one measurement by another and then multiplying by 100 to arrive at a percentage (see Platymeric Index in Chapter 3). In the past, these statistics figured prominently in sex assignment, especially when derived from pelvic measurements, due to the simplicity of their calculation. One of the best studied indexes was formulated in 1948 by Sherwood Washburn and termed the ischium-pubic index. This is calculated by dividing the length of the pubic bone, measured from within the acetabulum to the superior-medial surface of the pubic face (distance 'A' in left image of Figure 4.1B), divided by the length of the ischium, measured from the same point within the acetabulum to the most distant point on the ischium (distance 'B' in left image of Figure 4.1B), and multiplying by 100. Washburn showed that, for Euro-Americans (Whites) and Afro-Americans (Blacks), pelvises with index values under 84% are male while those over 94% are female. Although this statistic has been studied on many other populations, the difficulty in locating the point in the acetabulum reduces the use of this statistic in sex estimation. More significantly, the availability of computers in the 1950s and 1960s allowed for more complex methods (i.e., those requiring intensive computations on large amounts of data) to be applied to the problem of sexing. These newer methods, often called morphometrics, had the added advantage of being more accurate than simple indexes.

The most commonly used of these morphometric methods is discriminant function analysis (DFA) described in Chapter 1. This procedure calculates group membership using two or more measurements. Measurements from a skeleton of unknown sex are entered into a function, and if it exceeds a certain value (the sectioning point), it is considered to be of one sex, and if less it is the other sex. Probably the most famous application of this procedure in human skeletal biology was a 1963 study that used measurements of the skull to estimate sex. In this study, nine measurements of 408 skulls of known sex were taken and used to calculate 21 (later reduced to 14) different discriminant functions, using as few as 4 and as many as 8 measurements. (Multiple functions were calculated because skulls, especially those from prehistoric sites, are often incomplete and cannot be measured for all measurements.) As an example, suppose the skull of an Afro-American (Black) has the following measurements (in mm): maximum length=185, maximum breadth=135, distance across both zygomatic arches=128, height of the mastoid process=35. These values can be entered into the following function given in this study:

$$DF = (2.111*185) + (1*135) + (4.936*128) + (8.037*35) \approx 391 + 135 + 632 + 281 = 1439.$$

The quantities 2.111, 1, 4.936, 8.037 are called coefficients, and are multiplied by the cranial measurements to calculate the function value (DF). This value is compared to the sectioning point of 1388, with function values below that amount considered female while those above are considered male. Since the above function calculated a value greater than the sectioning point, this skull is probably male. All discriminant functions for estimating sex from bone measurements work the same as in this example. Measurements are multiplied by coefficients, the results of which are added together, and the final value compared to a sectioning point. Studies involving discriminant function analysis of human skulls show an accuracy between 80% and 90%, depending on the function used.

## OTHER METHODS

In addition to the visual and metric methods just described, sex can be estimated from biochemicals retrieved from bones and teeth. The most obvious biochemical useful for this purpose is DNA. Samples of bone can be analyzed for the presence of Y-chromosomes. (As a reminder, females have two X-chromosomes, while males have one X- and one Y-chromosome.) The absence of these chromosomes indicate that the bone is from a female, while their presence indicates male. Although this sounds simple, the process is complex and requires samples that have not been contaminated or degraded to the point that chromosomes cannot be identified.

Another such biochemical is amelogenin. This protein is involved in the formation of tooth enamel and can be retrieved from tissues in the pulp cavity of suitable teeth. There are two basic variants, one produced by a gene on the X-chromosome, and the other produced on the Y-chromosome. If both the X and Y variants are present in a specimen, then male sex is indicated; if only the X variant is present, the specimen is from a female. Interestingly, despite the apparent accuracy of this method, an assignment to male or female is not 100% correct (error rates close to 2% have been reported) due to factors such as biochemical degradation, contamination, measurement error, and genetic variation.

Geometric morphometrics is another method for estimating sex (among other uses). This is a more complex version of morphometric analysis that uses a shape component in addition to size differences. This method, when applied to the obturator foramen, was able to correctly sex 88.5% of male and 80.8% of female ossa coxae in a sample of 104 males and females. Even better results were found in a total of 64 sacroiliac joints, where 94.5% of individuals were correctly sexed.

## SEXING SUBADULTS

As mentioned above, the methods using traits and measurements of the pelvis, skull and other bones to estimate sex can only be used on adult remains (i.e., persons who were around 18 or more years of age at death). In subadults (i.e., individuals who died in infancy, childhood or early adolescence), these methods do not work very well, if at all. Although the same structures used in adult sexing have been studied in subadults, the results are rarely better than 50–50. For example, a wide greater sciatic notch (measured as greater than 90°) has a p=.556 probability of being female, and p=.444 of being male. Other characteristics of the ilium have been studied for this purpose, and the results are similarly discouraging so that sexing subadult bones is usually avoided.

## SUMMARY

1   The pelvis has the best visual indicators of sex, especially the three traits of the pubic bone: ventral arc, subpubic concavity, and medial border.

2    There are four visual traits of the skull that aid in estimating sex: supra-orbital ridges, nuchal crest, mastoid process, and mental eminence.

3    The metric lengths, widths, heights, and angles of various bones also can be individually used to estimate the sex of a skeleton.

4    Indexes, and more importantly, discriminant functions use multiple measurements to estimate sex.

5    Although very complex, there are a number of biochemical and geometric morphometric methods that can be used when sexing a skeleton.

6    Sexing subadults usually is not done using visual and metric measurements.

## FURTHER READING

For general information on sexing: WM Krogman and MY Işcan, *The Human Skeleton in Forensic Medicine*, 2nd ed. (Springfield: IL: CC Thomas & Sons, 1986), TD Stewart, *Essentials of Forensic Anthropology* (Springfield: IL: CC Thomas & Sons, 1979), WM Bass, *Human Osteology: A Laboratory and Field Manual*, 5th ed. (Springfield, MO: Missouri Archaeological Society, 2005), SN Byers and CA Juarez, *Introduction to Forensic Anthropology*, 6th edition (New York, NY: Routledge, 2023). For three traits of the pubic bone: TW Phenice (1969). A newly developed visual method of sexing the os pubis (*American Journal of Physical Anthropology*, 30:297–302, 1969). For the frequencies for pubic bone traits: AR Klales, SD Ousley, JM Vollner. A revised method of sexing the human innominate using Phenice's nonmetric traits and statistical methods (*American Journal of Physical Anthropology*, 149(1):104–114, 2012), MW Kenyhercz, AR Klales, KE Stull, KA McCormick, SJ Cole, Worldwide population variation in pelvic sexual dimorphism: A validation and recalibration of the Klales et al. method (*Forensic Science International*, 277: (259):e1–259.e8., 2017). For frequencies of skull traits: PL Walker, Sexing skulls using discriminant function analysis of visually assessed traits (*American Journal of Physical Anthropology*, 136(1):39–50, 2008). For morphometrics: AH Rosset al., *Geometric Morphometric Tools for the Classification of Human Skulls* (Report to the US Department of Justice, 2010).

# ESTIMATING AGE AT DEATH

The third part of the biological profile involves estimating the age-at-death of persons from their bones and teeth. There are a large number of these that are quite complex, so only the simple methods will be presented to reduce unnecessary detail. Age at death (AAD) methods can be divided into two basic types: those based on the developing skeleton, and those based on the deteriorating skeleton. Methods of the first type generally are used to estimate age of subadults; that is, infants or children who died from before birth to around 18 years of age. The second type helps with estimating AAD of adults; these are persons who died sometime around 18 years of age or older.

For the sake of simplicity, human age can be divided into seven stages: fetal (before birth), infant (birth to 3 years), child (3 to 12 years), adolescent (12 to 20 years), young adult (20 to 35 years), middle adult (35 to 50 years), and old adult (50 years and older). Notice the overlap in age ranges. This shows the inexact nature of estimating age from the skeleton. Skeletal biologists usually do not estimate an AAD to a single year but provide an age range of when a person died.

## SUBADULTS

Although there are many methods for estimating the age of subadults, they are all based on the developing and growing skeleton and dentition. Starting in the womb, bones begin to form (ossify) around the $6^{th}$ to $7^{th}$ week of pregnancy and continue to appear until birth when most bones are present but are incomplete and/or

DOI: 10.4324/9781003487944-5

made up of a number of segments. In infants and children, more bone segments appear that later join (fuse) together until they begin to resemble the mature bones they will become. This process of fusion continues through adolescence and into young adulthood when it stops in the middle to late 20s. While bones are forming (ossifying) and joining (fusing), teeth begin to form and grow and push (erupt) into the mouth. Starting with the milk (deciduous) teeth and followed by the permanent teeth, this process continues from infants to children to adolescents, where it stops in late adolescence. These topics will be discussed below and the schedule of each of the events will be presented.

OSSIFICATION AND EARLY GROWTH

Although bone forms within different tissues of the fetus (cartilage and mesenchymal tissue), it first appears in what are called centers of ossification where boney tissue begins to replace the original tissues. This replacement continues outwardly until they begin to resemble mature bones. Most bones have several centers of ossification that then fuse according to a rough schedule. It is the knowledge of the schedule of the appearance and growth of bone in the ossification centers, the schedule of fusion of the separate segments, and the increasing length of bones that allows for an age at death to be estimated for the unborn to newborn baby.

A simple example shows how this works. The radius (lower arm bone) ossifies from one primary ossification center and two secondary centers; the primary center ossifies the radial shaft (diaphysis) while the secondary centers ossify the upper (proximal) and lower (distal) ends (epiphyses). The shaft first appears as cartilage and begins to be replaced by bone starting around the 7th week of fetal life. This replacement proceeds from the center outwardly toward each end. The entire shaft is composed of bone at some unknown time but probably before 11 years of age. The proximal end (epiphysis) is made up of cartilage until around 5 years of age when the secondary ossification center begins to deposit bone in the cartilage; it is fully made up of bone by the time it joins (fuses) to the shaft by around 11.5 to 13 years of age in girls and 14 to 17 years of age in boys. Lastly, the secondary center of ossification in the distal epiphysis begins to deposit bone around 5 years of age and is fully ossified when it joins (fuses) to the shaft between 14 to 17 years of age in girls and 16 to 20 in boys. All bones follow a similar pattern, where bone forms from the various centers of ossification and grow outward until they reach an adult length. The bones formed from the primary and secondary centers then fuse to look like the mature forms of the bones as we know them.

FUSION OF PRIMARY OSSIFICATION CENTERS

The joining of primary centers to each other can be used to estimate AAD of subadults. The main skeletal elements that can be used for this function are the skull, mandible, atlas, and axis. In the skull, two types of unions occur. First, newborns exhibit gaps, called fontanelles, in the areas of bregma (the frontal fontanelle), pterion (the sphenoid fontanelle), asterion (the mastoid fontanelle), and lambda (the posterior fontanelle).

In addition to these gaps closing during life, the ossification centers of a number of cranial and postcranial bones unite during growth. This is true of the right and left halves of the frontal bone, which are divided from each other by the metopic suture at birth (see Figure 7.1D). Similarly, there are four parts of the occipital that fuse during development: the squamous part, the right and left lateral parts, and the basilar part. The squamous part is the majority of the bone and is made up of the back (posterior) and bottom (inferior) part of the bone, up to the back (posterior) of the foramen magnum (i.e., from lambda to the posterior foramen magnum). The lateral parts, one on the left side and one on the right, surround the rest of the foramen magnum in front (anterior) of the squamous and have the occipital condyles. The basilar part joins to the front of the right and left lateral parts and extends forward until it contacts the sphenoid bone. Additional fusions occur between the left and right posterior halves of the atlas as well as the anterior part (see Figure 2.6A). The mandible, which is divided into two halves while growing in the womb, also eventually fuses into one bone. Finally, the axis is composed of four elements: the body, dens, and right and left halves of the arch. The schedule by which these fontanelles close and primary centers join during growth is given in Table 5.1 and are useful only for persons under 10 years of age.

*Table 5.1* Times of Closure and Fusion in Early Childhood[1]

| Structure/Bone | Formation | Time of Closure or Fusion |
| --- | --- | --- |
| Fontanelles | Sphenoid and mastoid | Soon after birth |
| | Occipital | During first year |
| | Frontal | During second year |
| Mandible | Right and left halves | Completed by second year |
| Frontal | Right and left halves | In second year (remains open throughout life in as many as 10% of people) |
| Atlas | Union of posterior arches (posteriorly) | In third year |
| | Union of anterior to posterior arches | In sixth year |
| Axis | Dens, body, and both arches | In third and fourth years |
| Occipital | Squamous with lateral parts | In fifth year |
| | Lateral and basilar parts | In sixth year |

[1] Taken from Table 9.2 in Byers (2017). See Further Reading for the full citation.

## LONG BONE LENGTHS

The age at death of skeletal remains from unborn to newly born infants can be estimated from the lengths of their long bones. Although differences in size exist between subadults from different ancestral groups and sexes, lengths from around the 4th month after conception to the time of birth shows less variation among groups than later ages (e.g., childhood, adolescence, adulthood). Because differences in size accelerate after birth, the lengths of fetal, infant, and child long bones are only reasonably accurate indicators of age up until around seven years of age.

From the time of its first appearance in the 4th month after conception and up to birth, the length of the femur is a fairly reliable estimator of age. This is because there is very little variation in length for a given month after conception no matter the sex or ancestral group of the unborn child. For example, around the 6th prenatal month, the femur varies between 40 and 43 millimeters; this means that the longest femur (43 mm) is only around 7.5% longer than the shortest femur. This is in sharp contrast to where the difference between a femur from an adult man can be 25% (or more) longer than that of an adult woman. This small amount of difference for prenatal age makes it possible to get a reasonably accurate estimation of age from the femur up to around one month after birth. If the femur is not present but other long limb bones are, age can still be estimated since there is a close relationship between femur length and other limb bones. For example, if a 45 mm tibia is found, AAD can be estimated by knowing that this bone is around 89.9% of the length of the femur. This means that, if the femur were present, it would be $(45 \div .899 =)$ 50 mm which is the average length of a femur at around the 7$^{th}$ month after conception. This would be the estimated age of the person represented by the tibia.

After birth, the length of long limb bones becomes a less reliable estimator of age. Studies on skeletons of known age show that bone growth becomes more variable such that by 12 years of age the lengths of the long bones vary considerably from child to child. For example, a 60 mm femur indicates an age at death of between the 8$^{th}$ month after conception and birth; this is an age range of one month. However, by the time a femur is 180 mm, the age at death is 2.5 to 3.5 years, or an age range of one year. The increasing age range continues to get larger during growth such that when the bone is 340 mm, death occurred between 8.5 and 11.5 years of age, a range of three years. This increasingly poor indicator of age only gets worse through the rest of adolescence into young adulthood. Thus, bone length is only useful for age estimation in pre-born and early childhood.

TOOTH FORMATION AND ERUPTION

Although the formation and growth schedules described above can be used to estimate age at death of infants and children, the most commonly used method for estimating ADD in subadults is tooth formation and eruption. As described in Chapter 2, humans have two sets of teeth: the deciduous (often referred to as milk) teeth of infants and children, and permanent teeth of late childhood and adolescence (teenagers) and adults. During growth, teeth form in spaces inside the upper and lower jaws called crypts, following a pattern where the enamel covered tips of the cusps appear first, followed by the rest of the crown, after which the neck of the root develops with the final stage being the rest of the roots ending in the tips. This process occurs at different times for different teeth as illustrated in Figure 5.1. Looking at the 6th month of development, the tips of the cusps of the first permanent molar are visible as two small ovals on the far left. Moving to the right, the almost complete crowns of the two deciduous molars and canine are visible followed by the two deciduous incisors with roots partially developed. Above these last two teeth, the crowns of the permanent canine and 1st incisor are forming. In the second year of development, the process of tooth formation continues with the completed roots seen in the two deciduous incisors. Notice that once enough of the roots of the teeth have formed, they push through the bone to emerge (erupt) into the mouth as seen in the five deciduous teeth. The rest of Figure 5.1 reveals the complete formation of the five deciduous teeth first, followed by their slow replacement by the incompletely formed permanent teeth. Notice in years 6, 7 and 10, the roots of the remaining deciduous teeth are undergoing a reversal of formation; that is, they disappear (are resorbed) before the tooth is pushed out by the permanent teeth.

Looking at Figure 5.1, the age of subadults can be told by matching their dentition with the age that best fits the type (deciduous or permanent) of teeth that a present, the state of their development (from x-rays), and extent of their emergence. Several things need to be said about this figure. First, the figure only shows the upper dentition, not the lower teeth. This is because the lower teeth develop on a similar schedule to those of the upper dentition and are not pictured to avoid unnecessary detail. Second, Figure 5.1 is a simplified version of what is known about tooth formation and eruption in that it only has 10 age categories. Human tooth formation and eruption has been studied multiple times over the years both on living children as well as in skeletal collections mentioned in Chapter 1. This intensive study has led to a number of charts that have anywhere from 20 to over 30 stages. None of these are pictured here, again to avoid too much detail. Third, although all charts use stages of development, it must be

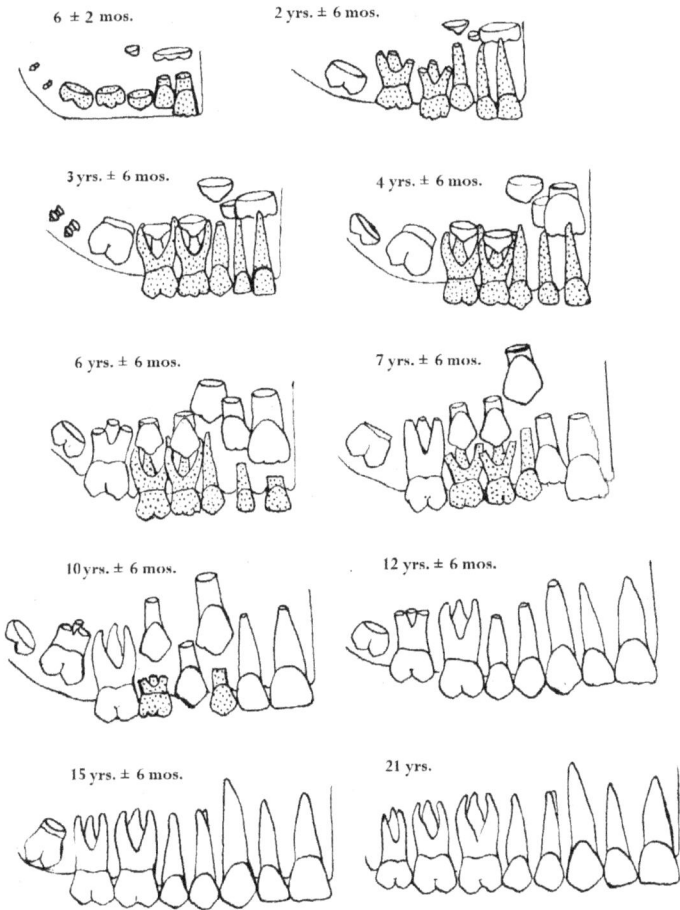

*Figure 5.1* Abbreviated Chart of Human Tooth Formation and Eruption at Various Ages. Stippled Teeth are Deciduous (Milk) Teeth

Source: Image from Figure 3.3A of *Digging Up Bones* by DR Brothwell (1981) Ithaca, NY: Cornell University Press. Used with permission.

remembered that tooth formation and eruption is a continuous process that proceeds in small, barely noticeable increases. Using stages based on age simplifies this complex process. Last, each age has a ± value attached to it (e.g., 3 yrs. ± 6 mos.). This means that there is a good chance that the actual AAD of a person with all deciduous teeth and only the first permanent molar erupted was probably between (6 yrs. − 6 mos. =) 5.5 years and (6 yrs. + 6 mos. =) 6.5 years. This shows the variability in age-at-death estimations using teeth, something that is seen no matter what tissues are used.

## EPIPHYSEAL UNION

As described earlier, most bones grow by the addition of boney (osseous) material on their ends. This starts after the primary centers of ossification have fused with each other and all of the original cartilage has transformed into bone. At this point, their ends (their joint surfaces) are not made of bone but are covered by cartilage that will eventually ossify. As growth continues, these ends (epiphyses) ossify into end plates that do not look like the articular ends they will become (Figure 5.2A1). As ossification and growth continues, the plates look more like an articular end and conform more closely to the ends of the shaft (Figure 5.2A2). As a person grows up, growth ceases, and the epiphyseal caps unite to their shaft (diaphysis), leaving a temporarily visible line where the two structures were once separated (Figure 5.2A3). Eventually, this line becomes completely erased (obliterated) by bone remodeling (Figure 5.2A4), leaving no trace of the original joint between the epiphysis and diaphysis.

There are many epiphyses in the human skeleton. All long limb bones have at least one on each end but often more. For example, the femur has four: one for the head, one for the condyles, one for each of the greater and lesser trochanters. Similarly, the humerus has three epiphyses: one for the head and both tubercles, one for the capitulum, lateral epicondyle and trochlea, and one for the medial epicondyle. Other bones can also have many epiphyses; the top 10 ribs have 3, 1 for the head and 2 on the tubercle: 1 for where it articulates with the transverse process of the vertebrae (called the articular part), and the other for the non-articular part of the tubercle. However, each metacarpal and metatarsal have only one, located at the distal end except for metacarpal 1 and metatarsal 1 where they are located on the proximal end. Also, the bottom 22 vertebrae have at least 5 epiphyses: one on the superior and inferior surfaces of the body, one on each transverse process, and one on the

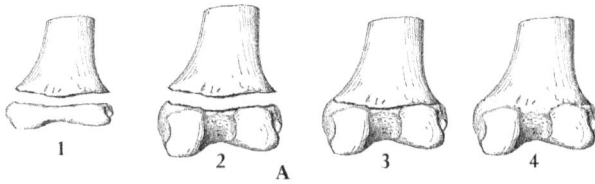

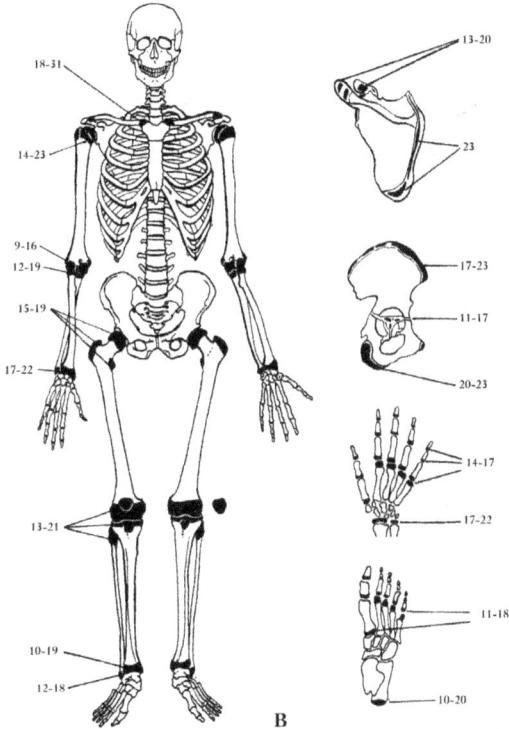

*Figure 5.2* Epiphyseal Union in Humans: (A) Stages of Union in the Distal
Femur, and (B) Location of Epiphyses and their Ages of Union in
the Human Skeleton

Source: Image B modified from Figure 3.4 of *Digging Up Bones* by DR
Brothwell (1981) Ithaca, NY: Cornell University Press. Used with permission.

tip of the spinous process. In addition, the lumbar vertebrae have two more, one on each mammillary process. Finally, the three bones of the hip bone (os coxae) fuse together in a manner similar to epiphyseal union as well as having three actual epiphyses.

Figure 5.2B shows where the epiphyses are in the human skeleton and their approximate time of fusion to the rest of their bones. This figure shows that epiphyseal union can be used to estimate the age at death of individuals who died in late childhood (9 to 12) through adolescence (13–18) and young adulthood (18–31). The times in Figure 5.2B are ranges of ages within which fusions occur. This means that bones where the line of fusion is very visible, as in Figure 5.2A3, are more likely from persons who died in the early part of this range while bones where the line is almost invisible or completely obliterated as in Figure 5.2A4, are from a person who died at the end, or after, the given time range. Also, although not shown in Figure 5.2B, some of the times of fusions are different for girls and boys, with girls' fusions usually occurring earlier than those of boys.

## ADULTS

Once the teeth are fully erupted and most bones have stopped growing, the skeleton begins to deteriorate; that is, it shows signs of wear-and-tear. It is the observation of areas of this wear-and-tear that led to the discovery of bone changes that help estimate the age of adult persons when they died. Although all bones and skeletal structures (e.g., pelvis, skull) deteriorate to one degree or another, two areas have been heavily studied in this respect: symphyseal surface (face) of the pubic bone, and auricular surface of the ilium. Each of these areas have a number of different features that change through time that were discovered while studying the skeletons of persons of known age at death. This section will describe these changes for each of the two areas as well as a short section on other age-related changes.

### FACE OF THE PUBIC BONE

One of the earliest skeletal areas to show age related changes is the face of the pubic bone, also called the symphyseal surface of the pubic bone. Changes to this surface follow a rough schedule according to age similar to those bone and tooth changes described for subadults. Although many researchers developed schedules and stages of changes, all looked at changes to the same parts of this area (labelled in Figure 5.3A): the bone of the pubic face (symphyseal surface), the front (ventral) and back (dorsal) margins, and the upper and lower borders (extremities). In addition, there are four other changes that occur on, or around, the symphyseal surface.

Starting with the surface of the pubic face, the bone shows a series of ripples (ridges) separated by valleys (furrows) from childhood to late adolescence that run across the surface giving the symphyseal face a saw-tooth appearance. As the person ages, the furrows fill in, usually starting from the back (posteriorly) and proceeding forward (anteriorly) until the face becomes flat with a grainy look to the bone. This is then replaced with fine-textured bone, which again becomes more grainy and finally becomes pitted and eaten away (eroded) in old age.

The front (ventral, anterior) edge (margin) of the pubic face develops two features over time: a rounded edge (bevel) and a bar of bone (rampart). In youth, the front of the pubic bone is at a right angle with the pubic face which makes the front edge somewhat sharp. However, as a person matures, the cortical bone of the front (ventral) surface of the pubis begins to invade the pubic face. This causes the sharp edge between the ventral margin and pubic surface to become blunted, until the margin curves (bevels) from the anterior surface of the pubis onto the pubic face. As time progresses, a ridge of bone appears on the ventral bevel, forming what has been called a rampart. This new bone causes the ventral margin to (again) form a right angle again between the front (anterior) surface of the pubis and the pubic face. This feature usually starts at the bottom (inferiorly) and begins to grow upward; at a slightly later time, it starts at the top (superiorly) and moves downward. Eventually, the two ramparts grow together, forming a single well-defined rounded bar of bone.

In addition to front (ventral) changes, the rear (dorsal) margin changes over time. Initially, the pubic surface is slightly curved from front to back. As the ridges break down and the furrows fill in, the bone of the dorsal pubic face builds up; this causes the surface to extend backward, forming a flat surface (plateau). This

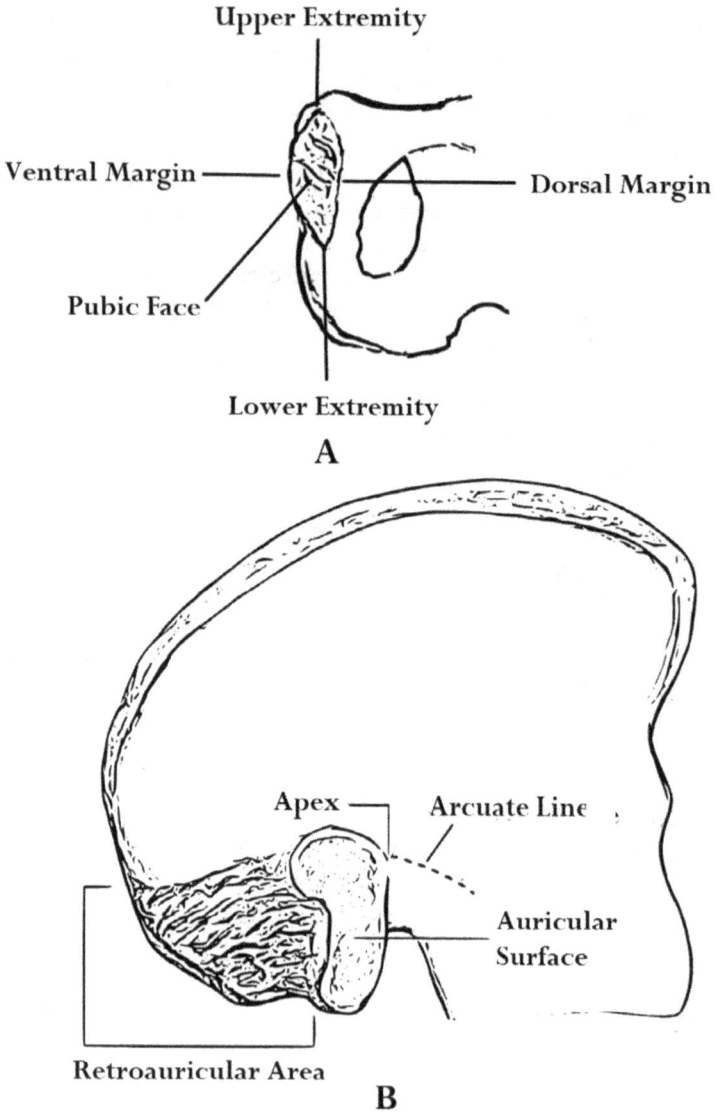

*Figure 5.3* Parts of the Pubic (Symphyseal) Face (A) and Auricular Area (B) that Change with Age

margin usually extends farther backward than the anterior surface of the pubic bone, causing its back (dorsal) edge to be fairly sharp.

The upper and lower edges of the pubic face, called extremities, also undergo changes. During youth, the extremities are not easy to see, because the bone of the pubic face blends with both the inferior and superior surfaces of the pubis. With aging, these edges become more defined, starting first with the lower extremity. Eventually, the upper extremity also becomes defined (although not as distinctly as the lower), and both extremities are distinguished easily by viewing the pubic face.

Four last osteological features that appear with increasing age are ossific nodules, the rim, lipping, and ligamentous outgrowths. Ossific nodules are "blobs" of bone that can be seen in early adulthood. These appear to aid in the formation of both the upper extremity and the superior part of the ventral rampart. Later, a rim is formed around the entire face of the symphyseal surface by the formation of both the upper and lower extremities, as well as the ventral and dorsal margins. This rim, which can be very distinct, usually is composed of cortical bone surrounding the rougher bone of the pubic face. The third feature is lipping. As age progresses, both the ventral and dorsal margins begin to curl outward, forming distinct lips. These structures are more visible on the dorsal surface, where they extend and thicken the edge already formed by the dorsal plateau. The last feature is ligamentous outgrowths that appear on the medial pubic face and are ossifications of the ligaments that join the right to the left pubic bone.

The changes of the different features described above correlate with different age-at-death ranges. Table 5.2 presents these AADs with their corresponding feature changes. The 'Avg' age in the first column is the average (mean) age of both males and females for the changes featured on that line of the table while range is the lowest and highest ages of both males and females. For example, on the fourth line of the table, avg = 37 and range = 26 to 70. This is the average age and age ranges of people whose pubic face has fine grained bone, defined borders and extremities, and a complete oval outline. Another feature seen in Table 5.2 is the size of the ranges for each AAD, which increases with age. For example, with average age of 19 (first line of table), the difference between the upper and lower limits of the range is (24–15=) 9 years, while it is (87–42=) 45

Table 5.2 Schedule of Changes in the Pubic Symphyseal Face[1]

| Average and Age Range | Symphyseal Face | Margins | | Extremities | | Miscellaneous |
| | | Ventral | Dorsal | Upper | Lower | |
| --- | --- | --- | --- | --- | --- | --- |
| Avg: 19 Range: 15–24 | Ridges and furrows | Beveling may have started | Undefined | Undefined | Undefined | Ossific nodules may be present |
| Avg: 24 Range: 19–40 | Ridges and furrows may be present | Rampart beginning to form | Undefined | Starting to be defined | Starting to be defined | Ossific nodules may present |
| Avg: 30 Range: 21–53 | Smooth or ridges and furrows still present | Rampart almost complete | Plateau complete | More defined | Almost defined | Ossific nodules may be fusing to form upper extremity and ventral border |
| Avg: 37 Range: 26–70 | Fine grained with remnants of ridges and furrows | Defined but gap can occur in upper part | Defined with possible lipping | Defined | Defined | Oval outline complete, possibly with rim; ligamentous outgrowths may occur inferiorly |
| Avg: 46 Range: 25–83 | No ridges and furrows; slightly depressed compared to rim | Defined with possible breakdown superiorly | Defined with moderate lipping | Defined | Defined | Rim complete; ligamentous outgrowths more prominent |
| Avg: 60 Range: 42–87 | Depression more common with pitting and porosity | Breaking down | Breaking down | Breaking down | Breaking down | Rim eroding; ligamentous outgrowths more prominent |

[1] Modified from data in Tables 9.5 and 9.6 in Byers (2017). See Further Reading for full citation.

years for the average age of 60 years (last line of table). This is typical of all methods for estimating age at death (as seen with subadult age methods); there is a decrease in accuracy with increasing age. One last thing to consider is what age a person is who has features from two lines on the table. In these cases, age may be estimated as half-way between the average ages on each line. Thus, if the pubic face has features of both avg=24 and avg=30, an age in the high 20's would be the best estimate for AAD.

AURICULAR SURFACE OF THE ILIUM

Another part of the skeleton that shows age changes is the back (posterior) of the ilium (see Figure 5.3B). As with other methods for estimating age, this area was studied on skeletons of known age at death, and three areas showed changes that occurred from young adults through middle adults to old adults. Interestingly, these areas change at approximately the same rate in men as in women, and in Whites and Blacks.

The three parts of the posterior ilium (Figure 5.3B) that help in estimating age are: the auricular surface, its apex, and the retro-auricular area. The auricular surface is the L-shaped part where the sacrum and ilium join (the sacroiliac joint); it is easily recognized not only because of its shape, but also because it is a roughened area that is distinctly different from the smoothness of the sur-rounding bone. The apex is the anterior–superior corner of the auricular surface, located where the arcuate line intersects the auricular surface. The retroauricular area is that section of the ilium that lies behind the (posterior to) auricular surface; this oddly shaped region varies in contour from flat to fairly wavy.

Within these three parts, five features change with age. The first feature is the transverse organization of the auricular surface; this refers to the manner in which the bone appears to be arranged across the joint. In youth, this organization takes the form of bil-lows that are similar to, but not as distinct as, the ridges on the pubic face. As on the pubic face, these begin to fill in over time, becoming less well defined, and eventually they are replaced by lines, called striae. Finally, all transverse organization, including striae, is erased.

The second feature that changes with time is the texture of the bone of the auricular surface. This bone starts out granular in youth but becomes coarser with age; this coarseness of the bone is described as having the same roughness as fine sandpaper. As time passes, this gives way to dense bone that is similar to, but not as smooth as, compact bone of the rest of the ilium. Eventually, this smoothness is lost as the bone degenerates. The third feature is the occurrence of pores in the auricular surface. Initially, these are small (called microporosity) but they become larger (called macroporosity). Some of the larger pores can be as much as 10 mm in diameter, although, generally, they are much smaller.

The fourth feature to change with age is the retroauricular area. The bone of this region starts out smooth and youthful looking before changes, but becomes rougher over time, and eventually is covered with boney bumps (osteophytes). These 'bumps' make the surface appear old, making it easy to distinguish from the smoothness of younger age. The final structure to undergo changes with time is the apex, which before activity begins is a thin crescent border that is almost sharp enough to cut the finger of a researcher. Over time, this feature thickens so that by old age it is several millimeters wide.

Table 5.3 shows the condition of each of these five features for various age ranges. By assessing the expression of each feature, age can be estimated from the row that best describes the bone being aged. For example, suppose that an os coxae is found with no billows but some striae composed of both granular and dense bone over the auricular surface and accompanied by some microporosity, some apical activity and slight coarsening of the bone in the retroauricular area. Comparison with Table 5.3 shows that three of the five features fall into the 40 to 44 age range (no billows but some striae, both granular and dense bone, slight apical activity, while three (slight apical and retroauricular activity, some microporosity) fall into the 35 to 39 range. Therefore, the best estimate of age at death for this individual would be the high 30s to low 40s.

OTHER METHODS

There are a number of other methods that can help estimate the age-at-death. One such method looks at the changes with age in the gap between the cranial bones that are separated by sutures. These

*Table 5.3* Changes in the Posterior Ilium by Age Range

| Age Range | Transverse Organization | Texture | Apical Activity | Retroauricular Area | Porosity |
|---|---|---|---|---|---|
| 20–24 | Billows | Very fine | None | None | None |
| 25–29 | Billows being replaced by striae | Slightly coarser | None | None | None |
| 30–34 | Less billowing, more striae | Distinctly coarser | None | Slight may be present | Some micro |
| 35–39 | Marked fewer billows and striae | Uniformly coarse | Slight activity | Slight | Slight micro |
| 40–44 | No billows; vague striae | Transition from granular to dense | Slight | Slight to moderate | Micro, maybe macro |
| 45–49 | None | Dense bone | Slight to moderate | Moderate | Little or no macro |
| 50–60 | None (surface irregular) | Dense bone | Marked | Moderate to marked | Macro present |
| 60+ | None | Destruction of bone | Marked | Marked with osteophytes | Macro |

[1] From Table 9.7 in Byers (2017). See Further Readings for the full citation.

gaps are particularly noticeable in young adults on the vault sutures (sagittal and lambdoid) where they can be 1–2 mm wide, but others of the braincase also show spaces between separate bones. (The sutures of the face are not used to estimate age.) These gaps get narrower with age and boney 'bridges' will develop across the suture lines. As time goes on, more such 'bridges' form and eventually the gap is erased entirely leaving what appears to be a line drawn on the braincase; in extreme cases, this line also disappears, and the evidence of a suture is completely erased (obliterated).

A simple method using suture closure to estimate AAD uses the amount of closure at six sites on the braincase: the midpoint of left-side of the lambdoid suture, the area around lambda, the area around obelion, the area about 1/3 of the way back from bregma to lambda, the area around bregma, and midway from bregma to left pterion. As a 1 cm section around each of these areas is examined, the amount of closure is assigned one of the following a scores: 0=open (no closure), 1=one boney bridge, up to 50% closure, 2=significant closure but a portion of the suture not fused, and 3=complete fusion and oblitera-tion. The scores for all six sites are added and the sum is compared to the values in Table 5.4 to estimate age-at-death.

Another method that is particularly useful for prehistoric and early historic remains is wear of the permanent molars. Early humans generally had coarse diets that contained grit that wore the chewing (occlusal) surface of teeth when food was eaten. As they aged, the enamel on teeth showed signs of wear that in turn led to exposure of the underlying dentin. As the persons aged, more and more of the enamel was worn away and more of the dentin was

*Table 5.4* Sum of Braincase Suture Closure Scores and Estimated Age[1]

| Score Sum | Average Age (Yrs.) | Age Range (Yrs.) |
|-----------|--------------------|------------------|
| 1 to 2    | 31                 | 18 to 44         |
| 3 to 6    | 34                 | 22 to 45         |
| 7 to 11   | 38                 | 27 to 44         |
| 12 to 15  | 45                 | 30 to 60         |
| 16 to 18  | 49                 | 35 to 59         |

[1] Combined from data in Table 9.10 and Figure 9.28 of Byers (2017). See Further Readings for full citation.

exposed until, in old age, only dentin existed on the occlusal surface. In one study, wear through the enamel of the first molar that exposed tiny amounts of the underlying dentin indicated an AAD of 17 to 25 years, while dentin being visible over around half of the chewing surface indicated 25 to 35 years of age. When there is little to no enamel left on the first molar, the person was 33 to 45 years of age at death. Finally, if only half of the molar teeth are left (the top half had worn away), an age-at-death of over 45 years is indicated. Wear on the second and third molars have similar age-at-death ranges but have less wear for the same ages because they erupt later than the first molar (the first molar erupts around 6 years of age, the second molar around 12 years of age, and the third in the late teens to mid-20s).

Another area where age can be estimated is the sternal ends of ribs; that is, the end of the ribs that attach to the sternum by the costal cartilage (see Chapter 2). The rib ends of young people have smooth bone with square, flat ends. As persons age, the bone becomes rough and pitted, and the ends become indented surrounded by a rim with sharp edges. Eventually, the rim becomes irregular as the costal cartilage transforms into bone (ossifies) such that long 'fingers' of bone jut from the rib ends. Research on the association of these changes with age developed tables of age versus changes similar to Tables 5.2, 5.3, and 5.4 above. There are other methods for estimating age other than those described above but require thin sectioning of bone or teeth and microscopic analysis. These are not described here since they are beyond the scope of a book of this nature.

## SUMMARY

1   Age-at-death can be estimated from changes in the skeleton that occur during growth and during the deterioration that slowly occurs in adults.

2   Fusion of the ossification centers in the skull and first two vertebrae (Atlas, Axis) follow a rough schedule that can be used to estimate age in infants and early children.

3   Limb bone lengths can estimate age-at-death in early children.

4   Tooth formation and eruption is the most commonly used method to estimate age-at-death in children from 6 months to 21 years of age.

5   Since all bones have epiphyses that fuse to the main part of the bone, epiphyseal union can estimate age in children to young adults.
6   Changes to the pubic face (symphyseal surface) and auricular area of the ilium are the most commonly used methods for aging adults.
7   Cranial suture closure, tooth wear, and sternal rib end changes can help establish age-at-death in adults.

## FURTHER READING

For age in prenatal remains and fusion of primary ossification centers: TD Stewart, *Essentials of Forensic Anthropology* (Springfield, IL: Charles C Thomas, 1979). For age from bone lengths in children: JM Hoffman, Age estimations from diaphyseal lengths: Two months to twelve years (*Journal of Forensic Sciences*, 24:461–469, 1979). For age from tooth formation and eruption: S Alqahtani, M Hector, H Liversidge, The London Atlas of Human Tooth Development and Eruption (*American Journal of Physical Anthropology*, 142:481–490, 2010). Age from epiphyseal union: JE Buikstra and DH Ubelaker, *Standards for Data Collection from Human Skeletal Remains* (Fayetteville, AR: Arkansas Archeological Survey Research Series44, 1994).
For pubic face changes, auricular surface changes, suture closure, and sternal rib ends: SN Byers, *Introduction to Forensic Anthropology*, 5th edition (New York, NY: Routledge, 2017). For age from tooth wear: DR Brothwell, *Digging Up Bones*, 3rd edition (Ithaca, NY: Cornell University Press, 1981).

# CALCULATION OF LIVING HEIGHT

An estimation of the height of persons from their bones is the last of the 'big four' questions that most skeletal biologists would like answered. Like other components of the biological profile, almost all skeletal biologists want to know the living height, also called stature, of persons represented by their skeletons. There are two different types of methods that can be used to get this characteristic: anatomical methods and mathematical methods. In anatomical methods, the bones are laid out in their proper position from head to toe and then measured for an estimate of living height. Mathematical methods use bone lengths to calculate stature using ratios, averages and regression formulas to compute this value.

In the following five sections, the subject of estimating stature from skeletal remains of adult humans will be examined. First, the calculation of stature from full, or nearly full, skeletons will be presented along with a description of the problems encountered when using this method. Second, the more common technique of using the total length of long limb bones to estimate this characteristic will be described. Third, the use of other skeletal elements (e.g., hand and foot bones, vertebral columns) for this task will be presented, followed by the methods used to calculate stature from incomplete (broken) long bones. The final section deals with adjustments to living height estimates due to age, and bone shrinkage.

## FULL SKELETON METHODS

Finding the living height of persons from their bones seems on the surface to be very simple: layout the skeleton as it is in life and

DOI: 10.4324/9781003487944-6

measure from the top of the skull to the bottom of the feet. Unfortunately, several things make this 'anatomical' approach difficult to use. First, when human skeletons are found, say in a prehistoric site, they are rarely complete enough to do this. Usually, the small bones such as vertebrae or feet are missing, or the skull is incomplete due to the effects of long-term burial. However, even if enough of the skeleton is found, it is very difficult to lay bones out on a table in their proper position for measuring. For one thing, the skull would have to be placed on its occipital, with the eye orbits facing the ceiling. Since the back (posterior) of the skull is round, keeping it in that position for measuring is not very easy. Next comes assembling (articulating) the vertebral column while taking into account the three curves described in Chapter 2: cervical, thoracic, and lumbar. The atlas (first cervical vertebra) and axis (second cervical vertebra) would need to be placed above the table in something like clay since they attach to the skull at the foramen magnum, which would be around two inches above the tabletop. The remaining cervical vertebrae would angle down from the skull to the tabletop where the top (superior) thoracic vertebrae would be set. These latter vertebrae would contact the tabletop at around the 3$^{rd}$ or 4$^{th}$ thoracic vertebrae, meaning that from the skull to these bones, the vertebrae would have to be spaced above the table. Proceeding downward, the rest of the thoracic vertebrae would also be above the table all the way to the bottom (inferior) lumbar vertebrae which would not touch the tabletop at all. The part of the skeleton that does touch the table is the back (posterior) part of the iliac crests of both ossa coxae, meaning that the pelvis would have to be assembled and placed on the tabletop in the manner it would be in life. Again, the vertebrae would have to be placed above the table so that they would end against the top of the sacrum, which is also above the tabletop. Next the head of at least one of the femurs would have to be placed in the hip socket (acetabulum) which would be above the tabletop. From that point, the femur and tibia can lay on the table in contact with each other. Finally, the two largest ankle bones, the talus and calcaneus, must be held together as they are in life and placed facing upward and the bottom of the tibia (talar facet of the tibia) would contact the top of the talus (the trochlea of the talus). These two joint surfaces (articular surfaces) would not contact the tabletop but would be about an inch above it. Even if this last task can be completed, the bones would have to be spaced apart as they are in life due to cartilage and intervertebral discs between the joint surfaces. This is particularly important with the vertebrae where the intervertebral disks add quite a bit to overall stature. Needless to say, this method is too difficult to regularly use to reconstruct the stature of persons from their bones.

What skeletal biologists have done is recognize that stature is made up of the heights and lengths of five parts of the skeleton: the height of the skull, the thicknesses of the bodies of the vertebrae, the height of the first segment of the sacrum, the length of the lower limbs, and the height of the talus and calcaneus when held together as they are in life. The sum of these values as well as an added amount for soft tissue (e.g., cartilage) is a reasonably accurate living height of a person. Figure 6.1 illustrates these measurements. Starting at the top, skull height (Figure 6.1A) is the distance from basion to bregma. Next, the heights of the vertebrae are taken from the upper (superior) to the lower (inferior) surfaces of the bodies (Figure 6.1B); except for the atlas, all cervical, thoracic, and lumbar vertebrae are measured as well as the height of the first segment of the sacrum. The measurement of the femur is its bicondylar length (Figure 6.1C) which is the distance from the bottom (inferior) of the femoral condyles, when they are both placed against the end of a flat surface, to the top of the femoral head. The length of the tibia excludes the intercondylar eminence (Figure 6.1D) but does include the malleolus, and finally, ankle height (Figure 6.1E) is the distance from the upper (superior) surface of the talus to the lower (inferior) surface of the calcaneus when they are held in natural articulation. After summing all these measurements, an amount for soft tissue thicknesses needs to be added. This includes the thicknesses of the skin of the scalp and bottom of the feet, the discs between each vertebra, the cartilaginous lining of the hip socket (acetabulum) and head of the femur as well as the cartilage between the femur and tibia (knee) and the tibia and talus (ankle). Research has shown that, if the sum of the bone measurements is 153.5 cm or less, the total of all of these soft tissues is around 12.2 cm in males and 11.2 cm in females. For bone sums of 165.5 cm and above, this amount is 14.0 cm and 12.9 cm for males and females, respectively, with sums between

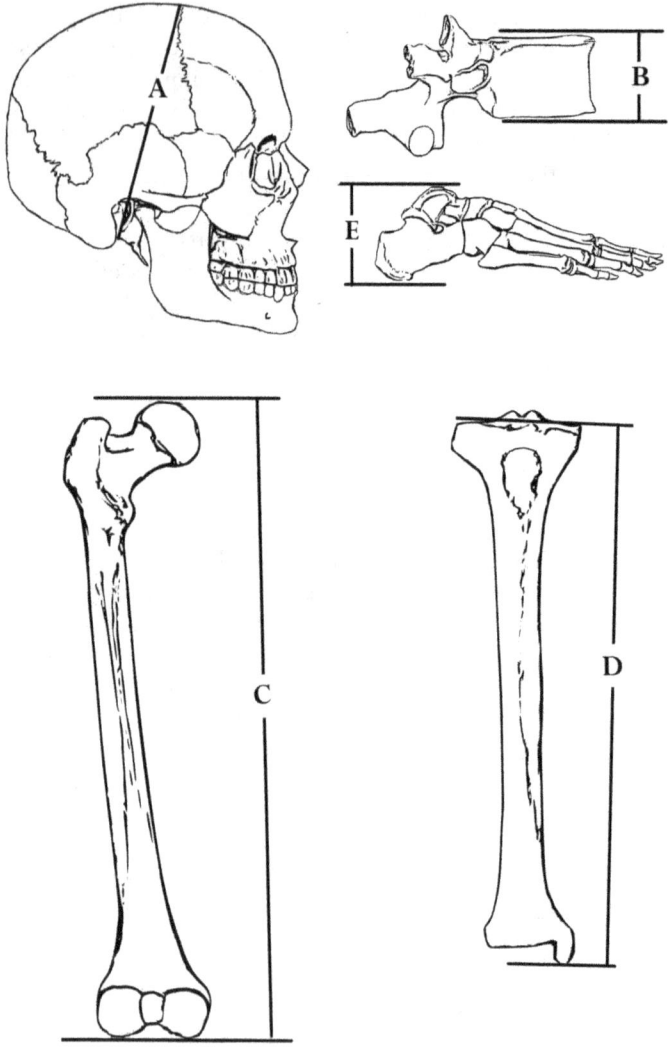

*Figure 6.1* Measurements Used in Full Skeleton Methods (See Text for Description)

these two amounts being 12.8 cm for males and 11.8 cm for females. Research has shown that this method of estimating living height is correct to within 4.5 cm (about 1 and ¾ inches) of actual height in 95% of persons.

## LONG LIMB BONES

Although adding the skeletal elements gives a pretty accurate estimate of stature, all of the bones must be present to use this method. This is often not the case, especially when prehistoric burials are excavated. Thus, a popular method for estimating height from the skeleton uses the total lengths of long limb bones; these are the lengths of the humerus, ulna, radius, femur, and fibula measured from the tops to the bottoms of these bones. (The length of the tibia is the same as that in Figure 6.1D.) The use of these lengths is based on the observation that tall people have long arms and legs, while short people have short limbs. This association between body segments has led to the development of ratios between bone length and stature, tables showing average statures for known bone lengths, and formulas (regression equations) that calculate living height from limb bone lengths. These three methods are collectively called mathematical methods.

Researchers in the 1800s to early 1900s measured the lengths of cadavers (usually dissecting room bodies) to arrive at an estimate of living height. They then measured lengths of different long limb bones and developed ratios of stature to bone length. For example, it was found that the humerus is approximately 20% of total living height, leading to a well-known (although inexact) method of multiplying the length of that bone by five to arrive at an estimate of stature. Other workers averaged the lengths of cadavers for a given bone length to calculate living height; for example, one such study found that for a 31.8 cm fibula, the average stature of persons in life was 153.0 cm (about 5 feet). Although many studies used these methods, advances in mathematics and statistics showed that regression equations were superior in calculating stature from limb bone lengths. These formulas are fairly simple in that the limb bone length is multiplied by a number (called a coefficient) and added to another number (called a constant) to arrive at an estimate of stature. Unfortunately, the calculation of the two numbers is very complex and tedious (when done by hand); when stature was calculated from the lengths of limbs on a little over 100 people in the early 1900s, it took several years of concentrated work to calculate the coefficient and constant. This hampered early workers from using this superior method, particularly on large samples. It was only when computers became more available for use (i.e., after around 1950) that these computations could be easily made on large numbers of measurements so that regression formulas now are used more often than the ratios and averages created by earlier workers.

The earliest stature formulas were calculated on known stature and lengths of long limb bones of over 5000 White and Black males and females from a collection of skeletons of persons of known height and from American servicemen killed in World War II and the Korean War. The Korean War sample included males not only of White and Black ancestry, but also Asians, Hispanics, and Puerto Ricans. Researchers discovered that the most accurate stature estimates were obtained when separate formulas were calculated for each sex and for each ancestral group. The measurements of the total lengths of the long limb bones were entered into a computer program that calculated formulas (regression formulas) that use bone length to estimate stature.

As time went on, new research has shown that the relationship between bone length and stature has changed in people for reasons that are not understood. The earliest formulas were calculated on persons born sometime in the mid-1800s to the early 1900s. However, modern skeletal collections contain people usually born in the middle 1900s and later. When the relationship between stature and bone lengths were compared between these groups, they differed significantly. Thus, newer formulas, based on modern populations, had to be calculated and are now used more than the earlier formulas.

A complete listing of all of these formulas is beyond the purpose of this book, however, Table 6.1 presents some of the formulas currently used by skeletal biologists. A few examples will help in showing how these equations are used and what other information they provide on people represented only by limb bones. From the table, the formula used to calculate stature from the total length of the tibia in Black females is: stature = (2.855 × Tibia Length) + 58.20. Thus, for a 35.0 cm tibia, the living height is: stature = 58.2 + (2.855 ×

*Table 6.1* Stature Reconstruction Formulas for Different Sexes and Ancestral Groups

| Formula | SE | Formula | SE |
|---|---|---|---|
| White Males | | Black Males | |
| St = 3.574*Hum + 57.21 | 5.71 | St = 3.277 ★ Hum + 65.46 | 5.72 |
| St = 4.525 ★ Rad + 61.22 | 5.70 | St = 4.235 ★ Rad + 63.46 | 5.07 |
| St = 4.534 ★ Uln + 53.33 | 5.66 | St = 3.979 ★ Uln + 62.95 | 5.79 |
| St = 2.701 ★ Fem + 48.10 | 5.12 | St = 2.455 ★ Fem + 56.66 | 4.84 |
| St = 2.891 ★ Tib + 62.95 | 5.06 | St = 2.455 ★ Tib + 75.48 | 5.03 |
| St = 2.832 ★ Fib + 66.96 | 5.15 | St = 2.665 ★ Fib + 69.39 | 4.53 |
| White Females | | Black Females | |
| St = 2.534 ★ Hum + 86.62 | 5.32 | St = 3.785 ★ Hum + 47.35 | 4.56 |
| St = 3.530 ★ Rad + 83.29 | 4.81 | St = 3.781 ★ Rad + 75.20 | 5.01 |
| St = 3.346 ★ Uln + 82.82 | 4.51 | St = 3.285 ★ Uln + 80.70 | 4.18 |
| St = 2.624 ★ Fem + 49.26 | 3.58 | St = 2.449 ★ Fem + 54.86 | 4.34 |
| St = 2.351 ★ Tib + 80.11 | 4.26 | St = 2.855 ★ Tib + 58.20 | 3.83 |
| St = 2.487 ★ Fib + 76.51 | 4.16 | St = 2.993 ★ Fib + 55.83 | 4.29 |
| Asian Males | | Hispanic Males | |
| St = 2.68 ★ Hum + 83.19 | 4.25 | St = 2.92 ★ Hum + 73.94 | 4.24 |
| St = 3.54 ★ Rad + 82.00 | 4.60 | St = 3.55 ★ Rad + 80.71 | 4.04 |
| St = 3.48 ★ Uln + 77.45 | 4.66 | St = 3.56 ★ Uln + 74.56 | 4.05 |
| St = 2.15 ★ Fem + 72.57 | 3.80 | St = 2.44 ★ Fem + 58.67 | 2.99 |
| St = 2.40 ★ Fib + 80.56 | 3.24 | St = 2.50 ★ Fib + 75.44 | 3.52 |

[1] Shortened from Table 10.2 of Byers (2017). See Further Reading for full citation.

35.0) = 58.2 + 99.925 = 158.125 cm which rounds to about 158.1 cm, or around 5 foot 2 inches. (Metric measurements can be converted to American measures by dividing the number of centimeters by 2.54, which is the number of centimeters in an inch. Then the inches can be divided by 12 to get the number of feet with any remainder converted into inches.) What is nice about this (and other such formulas) is the simplicity; despite the complex calculations that are made to derive them, using these equations is simply multiplication followed by addition.

The calculation of stature given above implies that a Black woman with a 35 cm tibia was exactly 5 foot 2 inches tall in life. However, this is an estimate of the actual height of the person because it is based on data gathered from a sample of Black female tibial lengths and statures. In this sample, all Black females with 35 cm tibias are not 5 foot 2 inches; rather some are a little taller and some a little shorter. To help understand its accuracy, a margin of error can be attached to this estimate using what is called the standard error (SE) of the estimate that accompanies all regression formulas in Table 6.1. By multiplying the SE by 2, a margin of error above and below the estimate can be determined. In this example, the standard error of the estimate is SE = 3.83 cm, meaning that approximately 95% of Black females with this length of tibia fall between (3.83 × 2 =) 7.66 cm less than 158.1 cm statue and 7.66 cm more than 158.1 cm stature; that is: 150.44 cm to 165.76 cm. These estimates convert to approximately 4 foot 11 inches to 5 foot 5 inches. (These calculations are not strictly correct from the standpoint of statistical theory but are close enough for a book of this nature; interested readers can consult any statistical methods textbook for a more complete discussion of these calculations.)

Notice that there are no formulas for female Asians and Hispanics. This is because there have not been enough skeletons from these groups available for study. If stature is desired for one or the other of these groups, the male equation can be used and its result multiplied by .92; this is based on the finding that women (on the average) are about 92% of the size of men as discussed in Chapter 1. One last point about calculating stature from long bone lengths needs to be made. Sometimes there are several complete limb bones available for this calculation leaving a skeletal biologist with a choice as to which one to use. Some will measure all bones and average the calculated statures; their thinking is they are using all the data that is available to them. Although this seems like a good idea, the formula with the smallest standard error should be used as this will give the 'best' estimate of stature since its margin of error will be smallest.

Although this section deals only with calculating stature for Americans, researchers in other countries have used regression to calculate stature reconstruction formulas from long bones using local samples. Thus, there are regression equations relating stature to long limb bones from Portugal, South Asia (India), Mexico, South Africa, Thailand, Japan, and China. Also, as more skeletons become available for study, more specific formulas are being generated for populations here in the United States such as those for modern Southwestern Native Americans. This trend of more formulas from more places in the world will almost certainly continue since it's been discovered that bone measurements from radiographs (X-rays) and CT-scans on the living can be compared to the stature of the persons to generate equations. This eliminates the need to remove the soft tissue and measure the bones from the few bodies that become available for this type of research.

## OTHER BONES

As might be expected, other bones and structures can be used to calculate stature since the principle that tall people have long arms and legs also applies to hands and feet, vertebral columns, and even skulls. Any one of these can be used to estimate living height, but all of their estimates have greater inaccuracy (i.e., larger SEs) than the methods described above. Of the other bones, three have been studied enough such there are regression formulas using them to generate stature estimates for people of the United States: hand bones (metacarpals), foot bones (metatarsals), and the vertebral column. The formulas for metacarpals, metatarsals and the vertebral column have been calculated using diverse samples so there are separate formulas for White males and females, and Black males and females. The hand and foot bone equations use the lengths of these bones while the equations using the vertebral column are based on the lengths of the cervical, thoracic, and lumbar vertebral segments in the living; also, they were calculated on articulated vertebral columns, so they include the thicknesses of the intervertebral disks. All of these formulas have large SEs making stature estimates calculated from them fairly inaccurate. For example, the SE for White female stature from the length of the fifth metatarsal is around 65 mm, or about 2.5 inches. Thus, (2.5 × 2 =) 5 inches should be added and subtracted from the stature, which means that the person's actual living height is somewhere within a 10 inch range. Using the formula for stature from the femur of that same female, the calculated living height of the person would fall somewhere in a 5.5 inch range; about half that of the metatarsal. All of the other equations calculated from these bones have similar SEs meaning they have similar levels of inaccuracy. However, if a stature is desired and there are no long limb bones available, these formulas can be used. Again, these formulas are based on American samples, but researchers in other countries have developed equations of their own for various bones (metatarsals in Portugal) and structures (pelvis in Japan).

## PARTIAL LONG LIMB BONES

Sometimes bones are broken and incomplete so that a whole bone cannot be measured for a stature estimation. In these cases, partial limb bones can be used for this purpose by knowing what percentage of the bone is present and using that to reconstruct its total length. Then this total length can be entered into the appropriate long bone equation to get a living height. Although seemingly easy, measuring partial long bones is quite challenging since they have few features, especially on the shaft, that can act as easy to find endpoints for measurements. Figure 6.2 shows the different points that have been used as endpoints in four of the six long limb bones (the ulna and fibula have not been similarly studied).

To understand how this works, look at the femur. Notice that from the top (proximal) end (labelled '1' on Figure 6.2), the first feature that has been used as an endpoint is the center of the lesser trochanter (labelled '2'). In females, this segment from (1) to (2) is about 16.48% of the total length of the bone. Although this first endpoint is relatively easy to find, notice that the next endpoint (3) is where the ridge that runs down the back (posterior) side of this bone (linea aspera) separates into two lines (medial and lateral supracondylar lines). This point is not very distinct which means that measurements from any of the endpoints to there are prone to error. The average percentage that the distance from (2) to (3) for female femurs is 59.73%. The next endpoint (4) is the top (superior) edge of the intercondylar fossa, with the last point (5) being the bottom (distal) most point on the medial condyle. The distance from (3) to (4) is 15.48% of the total length, and 8.41% for (4) to (5) in females. Percentages such as these have been worked out for both sexes for the femur, tibia, and humerus but only for male radii.

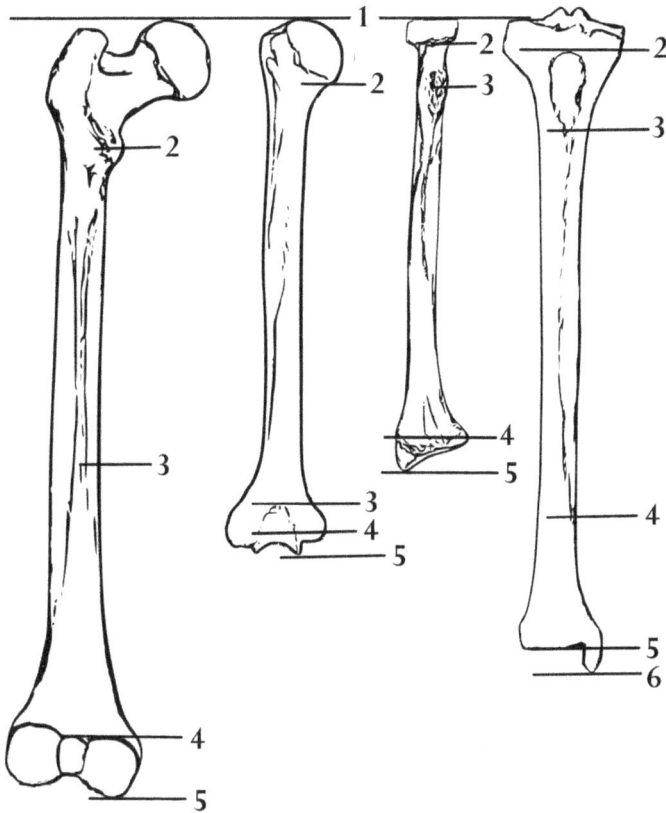

*Figure 6.2* Bone Segments Used to Reconstruct Bone Length

These percentages can be used to calculate total bone length in the following way. Suppose that an upper (proximal) portion of a female femur is found such that only the first two segments are present for measurement; that is the distance from the top (proximal) surface of the head (1) to where the supracondylar lines separate from the linea aspera (3). This means there is 16.48% of the bone from (1) to (2) and 59.73% from (2) to (3) of the bone present. If, say, the section is 28.9 cm, then the length of the femur is: femur length = $(28.9 \times 100) \div (16.48 + 59.73) = 2890 \div 76.21 = 37.9$ cm, or about 14 inches. Similarly, if the head of a female humerus is discovered and the length of the first segment is 3.0 cm, the total length of the bone would be (from the female formula): humerus length = (3.0 ★ 100) ÷ 10.77 = 300 ÷ 10.77 = 27.855 cm or about 11 inches. Unfortunately, statures derived from fragmented long bones are less accurate than those seen in almost all other bones (e.g., metacarpals, vertebral column) used to reconstruct stature. This is because the estimates of long bone length are inaccurate and have large margins of error. Add to that the SEs of the stature reconstruction formulas, and an estimation of living height from fragments of long bones result in very inaccurate estimates of stature.

## FACTORS AFFECTING STATURE

One factor that affects stature estimates is the age of a person. Most people have noticed that people get shorter when they get older, particularly when they are older than 60 years of age. This appears to be due to compression of the soft tissue between bone joints, especially the disks between the vertebrae. Although thinning of the cartilage of the ankle, knee, and hip play a role, the intervertebral disks are responsible for the greatest part of stature decrease due to their thickness. (For example, the disks between the lumbar vertebrae vary from 7 mm to 10 mm in thickness; this means that around 40 mm, or about 1.5 inches, of stature is due to the thickness of just the lumbar vertebral disks.) Studies of this loss in stature show that it is first noticeable around 45 years of age and speeds up over time. On the average, men loose about 3 mm or about an eighth of an inch by the time they reach 50 years of age. By the time they are 60 years of age, they have lost around 7 mm (a quarter of an inch), by 70 this amount averages around 16 mm, and after 70 they have lost around 32 mm (1.25 inches). Women fair better in that they lose height more slowly than men but by the time they are 70 years of age or older, they have lost an average of 34 mm (1.33 inches). By knowing the age at death of persons, living height estimates based on bone measurements can be decreased by the amount appropriate for their age.

Another factor that affects living height is the collapse of vertebral bodies due to fatigue fractures, especially among those suffering from osteoporosis (see Chapter 8 "Pathological Conditions"). This also leads to shorter vertebral columns, especially with the so-called "dowager's hump" in women. No studies have been done to correlate age and vertebral collapse so only the use of the full skeleton methods described above can result in a good estimate of living height in people suffering from these fractures.

A final factor is bone shrinkage. During life, bones are saturated with fats and other fluids that begin to evaporate after death, especially when all surrounding soft tissue has decayed away. This drying causes the bone to get smaller in all dimensions but especially in total length. Thus, shrinkage should be considered before bone lengths are entered into regression equations. The amount of shrinkage due to drying is related to the length of time that the bones were exposed and the relative humidity of the environment in which the remains decayed. Early estimates of amount of shrinkage showed a 2-mm shrinkage over a 10-month period in male and female long limb bones on a small sample. However, later studies using a larger sample both of Whites and Blacks showed 1.5% shrinkage in the femur. This amount is probably similar in all long limb bones and can be used to modify total length of fresh bones (Total Length *.985) before it is entered into a stature reconstruction formula.

## SUMMARY

1   The living height of a person, called stature, can be estimated by measuring all of the bones that make up height: skull, all vertebrae except the atlas, first segment of the sacrum, bicondylar length of the femur, total length of the tibia, and ankle height (talus and calcaneus).

2   Most statures from bones are calculated using regression formulas that relate bone length to living height.

3   Regression formulas calculate an estimate of stature but also provide statistics for a margin of error.

4   The lengths of the humerus, ulna, radius, femur, tibia, and fibula (collectively called the long limb bones) are most commonly used to calculate stature.

5   Many other bones (e.g., hand bones, foot bones, vertebral columns) and structures (e.g., skull height) can be, and have been, used to estimate height.

6   The total lengths of the femur, tibia, humerus, and radius can be calculated from their fragments, and this length can be entered into regression formulas to calculate stature.

7   There are several adjustments to stature that can be made: loss of height with age, collapse of vertebrae, and bone shrinkage after death.

## FURTHER READING

For general discussions of stature reconstruction from various bones, see relevant chapters: SN Byers and CA Juarez, *Introduction to Forensic Anthropology*, 6th edition (New York: Routledge, 2023), A Christensen, N Passalacqua, and E Bartelink, *Forensic Anthropology: Current Methods and Practice*, 2nd edition (London: Academic Press, 2019). For original description of the method for measuring all: G Fully, Un nouvelle méthode de détermination de la taille (*Annales Médecine Légale*, 35:266–273, 1956). For the earlier regression formulas based on persons born in the late 1800s and early 1900s: M Trotter, G Gleser, Estimation of stature from long bones of American Whites and Negroes (*American Journal of Physical Anthropology*, 10:463–514, 1952), M Trotter and G Gleser, A re-evaluation of estimation of stature based on measurements of stature taken during life and of long bones after death (*American Journal of Physical Anthropology*, 16:79–123, 1958). For newer stature formulas: R Wilson, N Herrmann, R Jantz. Evaluation of stature estimation using the Database for Forensic Anthropology (*Journal of Forensic Sciences*, 55 (3):684–694, 2010). For estimation of long bone length from fragments: DG Steele, Estimation of stature from fragments of long limb bones. In: T Dale Stewart, ed. *Personal Identification in Mass Disasters* (Washington, DC: National Museum of Natural History, Smithsonian Institution, 1970). For more information on Table 6.1: SN Byers, *Introduction to Forensic Anthropology*, 5th edition (New York: Routledge, 2017).

# SKELETAL ANOMALIES

Most skeletons have one or more 'odd features' that are not seen in the 'normal' skeleton described in Chapter 2. These oddities (also called anomalies, variants, or nonmetric traits) take many different forms including small, extra bones (called ossicles), unfused bone sections, enlarged places where muscles attach (enlarged tuberosities), single features that are multiple (multiple foramina), bridges of bone, and others. These anomalies are differences from what is considered normal and are caused by genetics, traumas, diseases, repeated strenuous activities or any combination of these factors.

Those 'oddities' that are controlled (at least partly) by genetics are unlike other skeletal characteristics described in this book in that they do not say much about the life of a person who has one or more of these variants. Rather, their main use is to show relationships between, and within, populations by calculating the percentage of persons who have each variant in an osteological collection. Research has shown that the percentage of these variants do not differ between males and females but do have different frequencies in different populations (e.g., Europeans vs. Chinese). These different percentages are due to a process called genetic drift, in which gene (trait) frequencies vary from generation to generation due to random chance. As human populations spread throughout the world, geographic distances prevented inter-marriage (and inter-breeding) between widely spaced groups. The result is that trait frequencies 'drift' apart and the longer that populations do not inter-breed, the more different the percentages become, and thus the greater the genetic separation. This means that their frequencies can be used to show the degree of relatedness

DOI: 10.4324/9781003487944-7

(called biodistance) between populations represented by skeletons, with similar percentages of these anomalies indicating close genetic relationships while different frequencies indicate more distant relationships.

Another cause of skeletal anomalies is occupational stress, and these do say something about the lives of the people who have them. Anomalies in this group are called markers of occupational stress (MOS) and have been found in people who work in different jobs that are so strenuous that the normal shape of their bones are modified during their lifetime. Generally, these changes are small such as extra bones inside the ear or facets due to bone-on-bone pressure. However, severe cases of scoliosis (S-shaped vertebral columns) and kyphosis (forwardly bent vertebral columns) can occur that are very large and very noticeable. Also, there is some overlap between MOSs and the pathological conditions described in the next chapter (Chapter 8 "Pathological Conditions").

In the following sections, the most common anomalies seen in the human skeleton will be described and illustrated. As stated above, these are: extra bones (accessory bones), extra (accessory) foramina, nonfusions (also called non-unions), and stress markers. In addition, Tables 7.1 through 7.6 will provide more complete lists of these anomalies and markers of occupation stress along with short descriptions. Thus, the reader should then have a good idea as to the large number of these variations and what they say about the lives of the people who have them.

## ACCESSORY BONES

The most commonly studied extra (accessory) bones of the skeleton are those that occur in the skull (see Figure 7.1A, B, and C). Notice that many of these are in the cranial sutures and hence are called sutural bones. They are also called Wormian bones after Ole Worm, a Danish physician (AD 1588–1654) who was one of the earliest scientists to describe these anomalies. Sutural bones are made from sections of two or three major cranial bones, such as the coronal ossicle that is made from parts of the frontal and one parietal (see Figure 7.1A and B) and the sagittal ossicle that is made up of parts of the left and right parietal bones (see Figure 7.1B) while the bregmatic bone can be made from both parietals and the

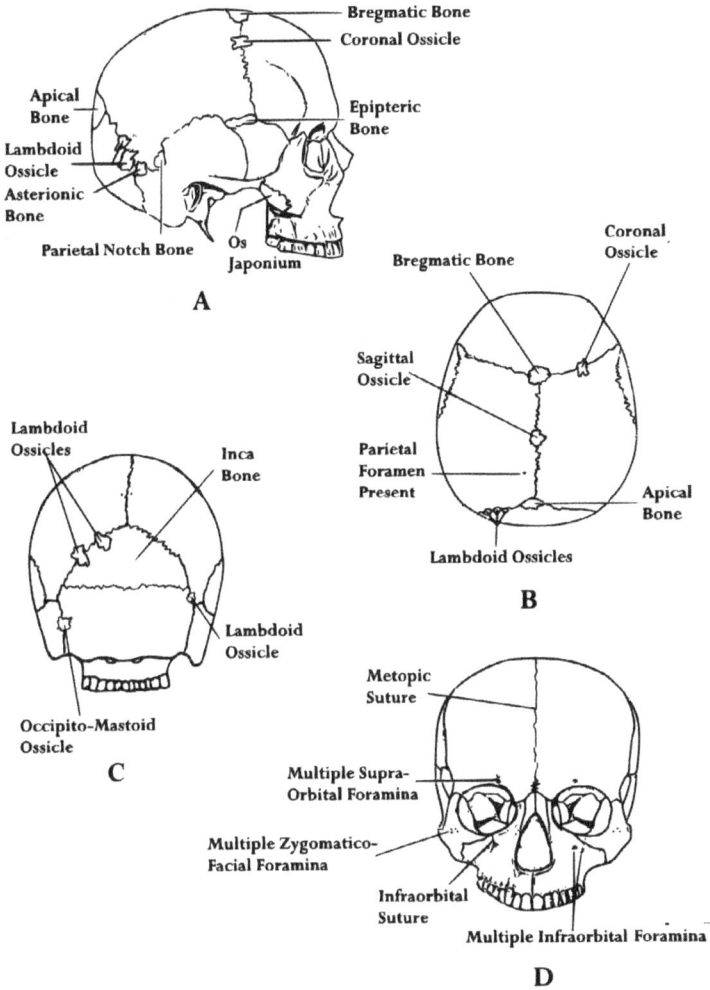

Figure 7.1. Accessory Bones and Multiple Foramina Commonly Found in the Human Skull (See Text for Description)

frontal (see Figure 7.1B) as well as just the two parietals. By contrast, the parietal notch bone can be made from a small segment of the parietal bone (see Figure 7.1A) or can be made from part of the temporal bone. Also notice that several of the bones are named after the landmark where they are located, such as the bregmatic bone at bregma and asterionic bone at asterion (see Figure 7.1A).

The occipital bone has several types of anomalous bones. The most common are the lambdoid ossicles that are made up of parts of the parietal and occipital bones (see Figure 7.1A, B and C); these appear in many prehistoric skulls from North America. Two accessory ossicles that are made up entirely of the occipital are the apical bone that appears at the apex of the lambdoid suture (see Figure 7.1A and B), and the much larger Inca bone, so called because they are common in prehistoric skulls from Peru, that takes up much of the occipital (see Figure 7.1C). This latter bone can appear as a single bone, a bone separated into right and left halves, or it can be divided into three parts. Two other sutural bones are the epipteric bone (see Figure 7.1A) and the occipitomastoid ossicle (see Figure 7.1C). The first can be made up of parts of the frontal, parietal, temporal, and sphenoid while that latter is made from parts of the occipital and temporal bones. One last accessory ossicle of the skull that is not embedded in a suture is os japonium (also spelled os japonicum) that is made from the lower (inferior) part of the zygomatic bone (see Figure 7.1A). These bones are so named because they appear most often in the skulls of Japanese people. One last oddity is a set of boney bridges that can occur spanning different foramens of the skull, dividing these normally single openings into two parts (not illustrated).

## ACCESSORY FORAMEN

In addition to accessory ossicles, the skull can have multiple (or missing) foramina that are usually single openings: supraorbital, zygomaticofacial, and infraorbital (see Figure 7.1D). Of these, the supraorbital foramen is most interesting in that it can be a single foramen, an incomplete foramen that looks more like a notch, or multiple foramina. Two other foramina are the parietal foramen that can be present or absent on the parietal bones (see Figure 7.1B)

*Table 7.1* Postcranial Nonmetric Traits[1]

| Nonmetric Trait | Description |
| --- | --- |
| Allen's fossa | A pit (fossa) on the anterior neck of the femur near the edge of the head |
| Poirier's facet | An extension of the articular surface of the femoral head down the anterior neck |
| Plaque formation | An extension of the articular surface of the femoral head down the anterior neck that is longer than Poirier's facet and can cover Allen's fossa |
| Hypotrochanteric fossa | A long pit (fossa) down the superior part of the posterior femoral shaft just below the less trochanter |
| Exostosis in trochanteric fossa | Spicules of bone in the pit (fossa) between the greater trochanter and the femoral head |
| Third trochanter | A small mound (tubercle) of bone near the base of the lesser trochanter on the lateral side of the femur |
| Supracondyloid process | A finger of bone pointing medially and inferiorly near the bottom of the humeral shaft |
| Septal aperture | An opening (aperture, foramen) through the boney wall that separates the olecranon and coronoid fossas of the humerus |
| Acetabular crease | A groove on the superior articular surface of the acetabulum |
| Pre-auricular sulcus | A groove on the superior boarder of the greater sciatic notch |
| Accessory sacral facets | A facet on the ilium posterior to the auricular surface; there can be a similar facet on the sacrum opposite it |
| Acromial articular facet | A facet on the inferior side of the tip of the acromial process of the scapula |
| Suprascapular foramen | A foramen formed by a bridge of bone across the scapular notch |
| Circumflex sulcus | Groove on the posterior and lateral border of the scapula |
| Vastus notch | Small notch on the superior lateral corner of the patella |
| Vastus fossa | Small pit (fossa) on the superior lateral corner of the patella anterior to the vastus notch |

| Nonmetric Trait | Description |
|---|---|
| Emarginate patella | Small notch on the superior lateral corner of the patella |
| Medial talar facet | Facet on the upper part of the neck of the talus |
| Lateral talar extension | Extension of the anterior trochlea onto the neck of the talus |
| Inferior talar articular surface | Extension of the articular surface of the talar head along the bottom of the neck |
| Anterior calcaneal facet double | Division of the calcaneal facet into two parts |
| Anterior calcaneal facet absent | Missing anterior calcaneal facet |
| Peroneal tubercle | Small bump (tubercle) on the anterior lateral side of the calcaneus |
| Atlas facet form | Division into two parts of the superior facet of the atlas |
| Posterior bridge | Boney bridge between the posterior edge of the superior facet and the posterior arch of the atlas |
| Lateral bridge | Boney bridge between the posterior edge of the superior facet and the transverse process of the atlas |
| Transverse foramen bipartite | Boney bridge across the transverse foramen of the $3^{rd}$ through $7^{th}$ cervical vertebrae dividing them into two halves |

[1] From Finnegan (1978). See Further Reading for the full citation.

and the mental foramen of the mandible (not illustrated) that is usually single but can be multiple.

## POSTCRANIAL ANOMALIES

Although cranial oddities have been studied the most, there are a large number of anomalies of the postcranial skeleton that, like those of the skull, are (at least) partially under genetic control. Whereas cranial oddities are mainly accessory bones and multiple foramina, there is a greater diversity of the types of postcranial anomalies: extra pits (fossas), extra facets, extra openings (foramina, apertures), extra boney knobs (processes, tubercles), extra grooves (creases, sulci), and boney bridges. For example, Allen's fossa is an extra pit (fossa) that is located on the front (anterior) side of the neck of the femur near the articular surface of the head (see Figure 7.2A). Poirier's facet is an extension of the articular surface of the femoral head down on the neck (see Figure 7.2B). Toward the bottom (distal end) of the humerus, the wall of bone separating the olecranon and coronoid fossas can be missing, which forms an opening called the septal aperture (see Figure 7.2C). Just above that, some people have the supracondylar process, a thin finger of bone that points outward and downward (see Figure 7.2D). An example of an extra groove is seen in the upper part of the articular surface of the hip socket (acetabulum) and is called the acetabular crease (Figure 7.2E), while an example of a boney bridge can be seen across the transverse foramen of the cervical vertebrae that divides this foramen into two parts, making it a bipartite transverse foramen (see Figure 7.2F). Table 7.1 lists these and the other most commonly studied postcranial anomalies.

## NONFUSION ANOMALIES

Nonfusion anomalies are oddities where parts of bones that normally fuse remain separate into adulthood. Many of these are failures of centers of ossification to join during growth so that, what should be a single bone in adults, are separate sections of the bone. These oddities can appear both in the skull and most (if not all) of the postcranial bones; in other words, they can occur in any bone that forms from two or more centers of ossification. In this section, only the most commonly seen nonfusions will be described and illustrated. Interested readers should consult the sources at the back of this chapter for a more complete list, with descriptions, of these types of anomalies.

There are a number of nonfusions that appear to be at least partially under genetic control. The kneecap (patella) has one of these conditions, called a bipartite patella. This condition most usually manifests as a separate bone in the upper-outside (superior-lateral) corner (see Figure 7.2G); however, these notches can occur in the upper-inside (superior-medial) corner (see Figure 7.2H) as well as the tip (see Figure 7.2I). Bipartite patella occurs in about 3% of the population and occasionally causes pain. Sometimes the neural arches of

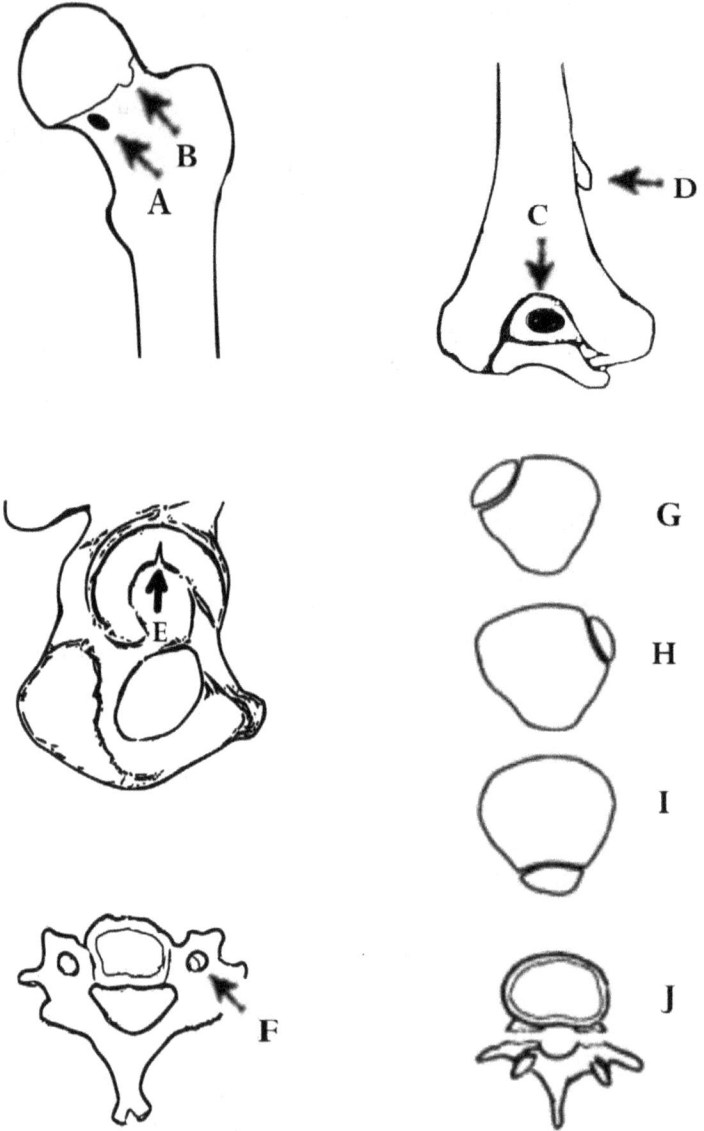

*Figure* 7.2 Postcranial Nonmetric Traits and Nonfusion Anomalies Commonly Found in the Human Skeleton (See Text for Description)

vertebrae do not attach to the body (see Figure 7.2J), a condition called spondylolysis which appears to be (at least partly) under genetic control but can be due to fracturing. There are several nonfusions of the acromion of the scapula (see Figure 7.3A) that may be under genetic control but also may be due to unhealed fracture in adults. Occasionally the styloid process of the ulna does not join with the rest of the bone in adults, resulting in what is called a persistent ulnar styloid ossicle (see Figure 7.3B). This small ossicle is not usually painful but when it is, surgical removal can cause the pain to disappear. On the talus, os trigonum is the name given to the posterior process when it does not join with the rest of the bone (see Figure 7.3C). As with the styloid ossicle, it too can cause pain but only rarely has to be surgically removed. Sometimes the metopic suture of the skull does not fuse by around eight years of age and remains open throughout life (see Figure 7.1D). This condition, called metopism, usually does not cause any harm to the person. Occasionally, a suture appears running from the lower border of the eye toward, and sometimes all the way to, the infraorbital foramen (see Figure 7.1D). What causes this suture is unknown since it cannot be due to a nonfusion between centers of ossification as the maxilla has only one such center. Sometimes the dens (odontoid process) of the 2nd cervical vertebra (axis) does not fuse and remains an ossicle throughout life (not illustrated) and there can be an accessory bone in hip socket called the os acetabuli (not illustrated) that is due to the nonfusion of one of the secondary centers of ossification to the rest of the os coxae.

Other nonfusions do not result in ossicles and can take many different forms. A common variant of the sternum is the sternal foramen. This is a hole through the body of the sternum (see Figure 7.3D) that, in some cases, looks a bit like a bullet hole. This oddity has been found in 2.5% to 13.8% of people and appears to be (at least) partially under genetic control since its frequency varies between populations. Another variant, called the tympanic dehiscence, is an opening on the front-bottom (anterioinferior) of the ear canal that occasionally occurs when this part of the temporal does not close during growth. Of all of the skull nonfusions, this has the greatest chance of causing a health issue in that its presence can cause balance and hearing problems in the person effected. Table 7.2 lists other common nonfusions found in the postcranial skeleton.

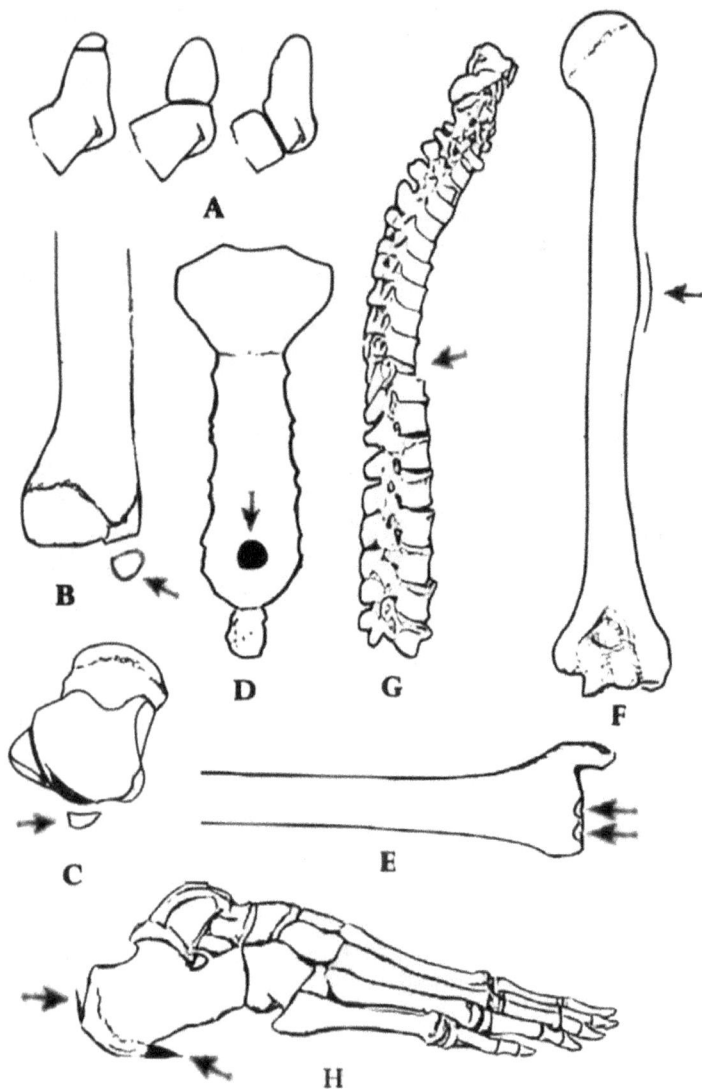

*Figure 7.3* Nonfusion Anomalies and Markers of Occupational Stress (MOS) Commonly Found in the Human Skeleton (See Text for Description)

*Table 7.2* Nonfusion Anomalies of the Postcranial Skeleton[1]

| Bone | Anomaly |
| --- | --- |
| Vertebral body | Nonfusion of right and left halves; nonfusion of anterior and posterior halves |
| Vertebral arch | Incomplete (hypoplasia) or missing (agenesis) of the pedicle |
| First Rib | Separation at the angle (congenital synchondrosis) |
| Sternum | Nonfusion of right and left sides (cleft sternum) |
| Clavicle | Nonfusion of left and right halves (congenital synchondrosis) |
| Humerus, Ulna, Radius | Nonfusion of proximal and distal halves (congenital synchondrosis) |
| Carpals, Metacarpals | Various nonfusions of the bodies, tubercles, and other structures |
| Os Coxae | Separation of ischium from pubis at the lower (inferior) junction (congenital synchondrosis) |
| Femur, Tibia, Fibula | Nonfusion of proximal and distal halves (congenital synchondrosis) |
| Tibia, Fibula | Nonfusion of malleolus |
| Tarsals, Metatarsals | Various nonfusions of the bodies, tubercles, and other structures |

[1] Taken from Table 15.1 in Byers (2017). See Further Reading for the full citation.

## OCCUPATIONAL STRESS MARKERS

Although the above-described variants may have genetic and/or traumatic/disease causes, there are many 'oddities' that are seen in people who work in activities that require the repeated use of strenuous muscular energy. Many heavy labor jobs (e.g., construction) and activities (e.g., furniture moving) can alter the shapes of bones of the persons who are engaged in them for a large part of their life. Enlarged areas of muscle attachment, regions of bone loss (erosion), and soft tissues that change to bone (ossifications) can indicate heavy use. The presence of these changes can help in understanding the lives of a people represented by their osteological remains. Although most of these markers of occupational stress (MOS) are likely to be seen only in third-world countries, there are a number of activities even in industrialized societies that are so strenuous that they can modify living bones.

Unfortunately, these oddities don't always indicate heavy activity since they can be seen as part of normal aging. Multiple studies have shown that modifications in the upper limbs are seen more often in older individuals than in younger individuals, which makes sense because the longer a person does a strenuous activity, the longer the bones have to modify their shapes. (Oddly, changes in the lower limbs have not been linked to old age; this may be due to the fact that the legs and feet have not been studied as much as the upper limbs.) However, these anomalies are also found in older individuals even though they were not engaged in occupations that involved heavy activities. This means that if they appear in the bones of a young person, it's reasonable to feel that the person probably engaged in an occupation indicated by the lesion. However, if these modifications appear in an older individual, their presence may be due to a strenuous occupation, or due simply to old age. Thus, when these oddities are encountered, the age at death of the person needs to be accounted for.

Markers of occupational stress can be divided into four basic types: changes to areas of muscle attachment, growths of thin fingers and/or ridges of bone (osteophytes), stress fractures, and other discrete markers. The most noticeable change to areas of where muscles (through tendons) attach to bones or where two bones attach to each other (through ligaments) is that the area of attachment (tuberosity, joint edge) gets larger and more rugged. This 'hypertrophy' of a tuberosity can be very noticeable and can sometimes affect the entire bone (although this is very rare). Strenuous activities can stimulate bone growth and cause thin spurs of bone (osteophytes) to grow at joints as well as rims of new bone to form around joints. (However, pathological conditions such as arthritis can also cause osteophytes; see Chapter 8.) The third type of MOS is stress fractures, such as what can happen to the bodies of vertebrae. Repeated strenuous activity can cause parts of bone to collapse due to numerous small fractures (microfractures). Also, these fractures can cause parts to separate from the rest of the bone (e.g., neural arch separates from the body). The last type is a collection of 'other' instances of MOS that are not easily categorized. An example of this is a pair of wear facets that appear on the front (anterior) part of the bottom (distal) end of the tibia (see Figure 7.3E). Such facets and other oddities are seen in other parts of the skeleton.

Table 7.3 Enlargement of Insertion Areas of Muscles from Activities or Occupations

| Bone or Structure | Area Enlarged | Action/Movement | Possible Activity/Occupation |
|---|---|---|---|
| Occipital | Tubercles around the superior nuchal line where various neck muscles attach | Hunching over | Miners; heavy construction |
| Temporal bone | Mastoid process | Pushing head forward against backward pressure | Tumpline use |
| Mandible | Enlargement of the pterygoid tuberocities | Jutting mandible forward and holding | Playing clarinet or other wood-wind |
| | Flaring at gonion | Pushing against backward pressure | Tumpline use |
| Clavicle | Costal tuberosity | Pressure on shoulders | Agriculturalists: ploughing, carry-ing weights on shoulders |
| Humerus | Lateral wall of the intertubercular groove and deltoid tuberosity | Paddling action with weights | Kayak hunters |
| | Medial epicondyle | Swinging forearm backwards | Javelin-throwing, golfing |
| Radius | Tuberosity | Carrying weights with elbows bent | Masons, bakers |
| | | Pulling arm backward against weight | Archers |
| Ulna | Supinator crest | Rotating forearm with heavy lifting | Fruit pickers, iron workers |
| Scapula, Ulna and Humerus | Areas where triceps muscle attaches | Flexing and extending arms | Net casting, woodcutting, black-smithing, baseball |

| Bone or Structure | Area Enlarged | Action/Movement | Possible Activity/ Occupation |
|---|---|---|---|
| Hands | Palmar insertions on the proximal phalanges | Squeezing thumb towards other fingers over extended period | Writers (holding pen/pencil), rowers (holding paddle or oar) |
| Femur | Gluteal tuberosity, linea aspera | Extending legs | Football players, horseback riders |
| Tibia | Tibial tuberosity | Holding legs in partial extension | Charioteers |
| Calcaneus | Heal spurs pointing vertically | Running, jogging | Runners, joggers |
| | Heal spurs pointing horizontally forward | Walking for long periods on hard surfaces | Police officers, floorwalkers |

[1] Reduced from Table 1 of Wilczak and Kennedy (1998). See Further Readings for the full citation.

One of the most common areas exhibiting a hypertrophic lesion is the tuberosity where the deltoid muscle attaches to the humerus (see Figure 7.3F). This has been seen in modern kayakers as well as Aleut kayakers of Alaska. In addition to the deltoid tuberosity, there are a large number of changes to areas of attachments of tendons and ligaments. The occipital bone can develop small tubercles in the tendons where various neck muscles attach around the superior nuchal line in miners who hunch over in small spaces. The mastoid process can enlarge and the area around gonion can flare outward when a person regularly carries heavy objects in a tumpline (a sling that crosses over the forehead and down the back to carry things). Playing the clarinet (and probably other woodwind instruments) can cause the pterygoid tuberocities of the mandible to enlarge, due to the forward movement of the lower jaw during this activity. An enlarged medial epicondyle of the humerus has been linked to javelin throwing, indicating that a similar lesion on a prehistoric man from Niger (Africa) may be due to spear throwing. Table 7.3 lists these and other such changes seen in the skeleton.

As people age, small spurs or ridges of bone project from an area that is normally smooth or flat. Called osteophytosis, these osteophytes occur around joint surfaces, especially in the vertebrae but also the shoulder, knee, and elbow. Vertebral osteophytes can be anywhere from small to large spurs of bone (see Figure 8.1F) that, over time, can join to form ridges. Unfortunately, understanding what causes these is hampered by that fact that they naturally occur with age; therefore, before trying to interpret activity and occupation of an individual from these osteophytes, the age of an individual at death must be accounted for. Joints where the bones move around (called diarthrodial joints) are most likely to develop ridges of bone rather than spurs of bone. Common areas with these lesions include the glenoid fossa of the scapula and the bones of the knee. Joints where the bones do not move much (called amphiarthrodial joints), such as the vertebrae, are more likely to develop spurs of bone that can sometimes get very large. Table 7.4 lists places in the skeleton where spurs and ridges occur and the activities that might have caused them to form. Some can be seen in the skeletons of people with occupations found in industrialized countries (e.g., using a pneumatic drill) while others are seen in skeletons of prehistoric people or people who engaged in activities found in third world counties (e.g., grain grinding with grinding stones).

*Table 7.4* Osteophytosis from Activities or Occupations

| Bone or Structure | Affected Area | Action/Movement | Possible Activity/ Occupation |
|---|---|---|---|
| Vertebrae | Bodies of C4, C5 and C6, C7 | Carrying heavy weight above or near head and neck | Fruit pickers |
| | All areas of cervical vertebrae | Pushing head forward against backward pressure | Tumpline use |
| Shoulder, elbow, hand | Joints and muscle markers of the shoulder arms and wrists | Forward movement of shoulders with bending and straightening of arms | Grinding grain with grinding stones (mano and metate) |
| Scapula | glenoid cavity (one side only) Both sides Glenoid cavity and wrist | Extending arm against pressure Same as above Repeated jolting activity | Archers Skin preparation (for clothing) Using a pneumatic drill |
| Elbow | Capitulum and radial head | Rotational movements | Grain grinding, spear throwing, kayaking |
| | Lateral epicondyle of humerus | Tugging on extended arm | Dog walking, ice fishing |
| Wrist | Joint between distal ulna and carpals | Flexing of hand | Skin scraping |
| Knee | Sides of femoral and tibial condyles | Kneeling | Grinding grain with grinding stones |
| Foot | First and second metatarsals | Kneeling | Grinding grain with grinding stones |

[1] Reduced from Table 2 of Wilczak and Kennedy (1998). See Further Readings for the full citation.

A third type of MOS involves fractures caused by stress from repeated and strenuous activities. Unfortunately, like osteophytosis, these fractures can occur in aging persons for many reasons unrelated to activity, especially among those suffering from osteoporosis. This is particularly true for compression fractures of the thoracic vertebrae (see Figure 7.3G). Therefore, the age at death of a person must be taken into account when evaluating these lesions as potential occupation-related markers. Table 7.5 lists the location and bone changes of those stress fractures for different occupations and activities.

The final type of MOS is collectively referred to as discrete markers of activity and can take many different forms in the skeleton: facets, grooves, deformities, tori, and accessory bones. Facets occur on two bones near a joint surface; the classic example of this type of lesion is the already described facets that occur on the anterior part of the distal tibia where it contacts the neck of the talus during squatting (see Figure 7.3E). In a similar manner, grooves can occur in areas characterized by smooth surfaces. A common location is on the occlusal areas of teeth when objects are held in the mouth (e.g.,

*Table 7.5* Fractures Associated with Occupations and Activities

| Bone Effected | Fracture | Activity/Occupation |
| --- | --- | --- |
| Lumbar vertebrae 1 thru 4 | Separation of neural arch from vertebral body | Heavy lifting |
| Vertebrae (middle and lower) | Anterior wedging due to compression fracture | Sledding, snowmobiling, parachuting |
| Cervical vertebrae | Fracture of arches due to neck rotation with compression | Carrying heavy loads on head |
| | Fracture of vertebrae | Milking cows |
| Radius | Bilateral stress fracture | Bakers, masons |
| Ulna | Chipping at trochlear notch with exostoses on medial surface | Baseball pitchers |
| Thumb | Transverse fracture | Rodeo or mechanical bull-riding |
| Fibulae | Stress fractures at top | Jumping from squat |
| Calcaneus | Exostoses and fractures | Impact of heel on ground (e.g., dismounting after horseback riding) |

[1] Reduced from Table 4 of Wilczak and Kennedy (1998). See Further Readings for the full citation.

Table 7.6 Discrete Markers of Activities or Occupations[1]

| Affected Area | Marker | Activity/Movement | Possible Group or Occupation |
|---|---|---|---|
| Palate and/or inner surface of mandible | Enlargement of bone into oral torus | Heavy chewing (e.g., hide preparation | Circumpolar peoples |
| Ear canal | Auditory exostoses (ossicles) | Repeated exposure to cold salty water | Divers |
| Kyphosis of spine | Forward angling of spine (due to collapse of anterior vertebral bodies) | Hunching while carrying heavy weight | Tailors, weavers, shoemakers, factory workers |
| Scoliosis of spine | Side-to-side angling of spine | Carrying burdens on one shoulder (can be congenital in origin) | Stone miners, heavy laborers |
| Vertebral bodies | Schmorl's node (depression in vertebral bodies) | Carrying heavy weights | Heavy laborers |
| Sternum | Manubrium fused to body Concave sternum | Heavy vertical force on shoulders Pressure against chest | Porters Shoemaker |
| Clavicle | Robusticity at acromial end | Carrying heavy weights with both arms while standing | Milkmen |
|  | Robusticity at sternal end | Hunching shoulders (e.g., while sewing) | Seamstress, tailor |
| Glenoid cavity | Facet on upper rim | Lifting heavy burdens overhead | Fruit pickers (citrus fruits) |
| Acromial process | Divided into two parts (bipartite) | Heavy shoulder loading | Archers (using long bows), Fruit pickers |

| Affected Area | Marker | Activity/Movement | Possible Group or Occupation |
|---|---|---|---|
| Ulna | Thickening (hypertrophy) of proximal half | Hammering action | Rodeo cowboys |
| Lower limbs and ankles | Facets on: upper femur, femoral condyles, anterior surface of distal tibia, and neck of the talus | Squatting | Grain grinders |
| Ischium | Roughened (craggy) tuberosity | Sitting for long time | Weavers, coachman, bargers, tailors |
| Foot | Extension of metatarsal–phalangeal articulation | Kneeling | Canoers |
| Front teeth | Antemortem loss | Holding something tightly in teeth (e.g., sled dog leads, fishing line) | Dog sled handlers, fishermen |
| Incisors | Dented edges | Holding things in teeth (e.g., tacks, nails) | Carpenters, upholsterers |
| Teeth | Grooves into chewing surface | Processing fibers | Fishermen, basketmakers |
| Premolars | Ellipsoid aperture | Clenching a rounded item in the premolar teeth | Pipe smokers |

[1] Shortened from Table 3 of Wilczak and Kennedy (1998). See Further Readings for the full citation.

nails by carpenters and shoemakers). The teeth can show a number of grooves and pits when they are used continuously for non-chewing activities. Deformities appear when a segment of bone is placed under repeated and (usually) high stress. An example is the slight flattening of the humeral head seen in persons who often raise their arms above their heads (e.g., citrus fruit pickers). Other such changes include tori (raised area of a bone) and accessory bones. These markers are found in most areas of the skeleton, emphasizing the need to inspect all bones during a skeletal analysis. One last discrete marker type is ossification of tendons where they attach to bones. A common place for this type is the foot since it can be under significant load even when walking without carrying heavy objects. Ossification of the tendon (Achilles tendon) that connects the calf muscle to the heal (calcaneus) and the plantar facia that connects to the front (anterior) of the calcaneus are two common areas where tendons become boney due to heavy use (see Figure 7.3H). Because these types of markers can have a genetic component, estimation of their relation to activity depends on knowledge of the population from which the skeletons were derived. Table 7.6 provides information on the major discrete markers for occupations and activities normally found in the United States and other industrialized countries.

## SUMMARY

1   The human skeleton has a number of oddities (anomalies) not seen in normal skeletons that take four forms: accessory ossicles, nonfusions, accessory foramina, and miscellaneous anomalies.

2   Small, extra bones occur in the skull, mainly within the sutures, that are (at least partially) under genetic control. Multiple foramina in the cranium also are (at least partially) under genetic control.

3   The nonfusion of normally complete bones are congenital conditions, which can appear in any part of the skeleton.

4   There are a number of other oddities, such as extra facets, extra foramina, grooves, and fingers of bone.

5   There are four types of lesions due to occupational stress: modifications to joints and areas of muscle insertion, osteo-phytosis, stress fractures, and discrete markers.

6    Markers of occupational stress (MOS) expand the knowledge on how people lived.

## FURTHER READING

For a list and description of the 84 known cranial variants, see G Hauser and G De Stafano, *Epigenetic Variants of the Human Skull* (Stuttgart: Schweizerbart, 1989). For many skeletal anomalies, nonfusions, and markers of occupational stress: SN Byers, *Introduction to Forensic Anthropology*, 5th edition (New York, NY: Routledge, 2017). For skeletal anomalies: J Buikstra and D Ubelaker, *Standards for Data Collection from Human Skeletal Remains* (Fayetteville, AR: Arkansas Archaeological Survey Research Series 44, 1994), E Barnes, *Atlas of Developmental Field Anomalies of the Human Skeleton: A Paleopathology Perspective* (Hoboken, NJ: John Wiley and Sons, 2012), M Finnegan, Non-metric variation of the infracranial skeleton (*Journal of Anatomy*, 125(1): 23–37, 1978). For markers of occupational stress: C Wilczak and KAR Kennedy, Mostly MOS: Technical aspects of identification of skeletal markers of occupational stress. In: KJ Reichs (ed.) *Forensic Osteology: Advances in the Identification of Human Remains*. 2nd ed. (Springfield, IL: Charles C Thomas, 1998).

# PATHOLOGICAL CONDITIONS

As is well known, humans are prone to many different types of diseases, also called pathological conditions, that have affected people throughout their entire history. Infections caused by bacteria, viruses, parasites, and fungi are common today and certainly were in the past. Malnutrition, trauma, congenital conditions, tumors, and arthritic degeneration are also present today as they were in the past. Additionally, teeth are affected by various bacteria that cause cavities (called caries) or plaque, which can cause abscesses and gum disease that can lead to bone resorption. This chapter will deal with some of the most common pathological conditions that affect the human skeleton and those that affect the dentition. As always, more detail is available in the books listed in the Further Reading section.

## SKELETAL PATHOLOGIES

Of the many diseases (pathological conditions) that affect people, only a few cause osteological changes that can be seen in the bones. These changes can be thought of as causing abnormal bone loss, abnormal bone gain, abnormal bone shape, and trauma (e.g., breaks, cuts, holes). Abnormal bone loss can be seen in the skeleton as destruction of bone, both of the surface (cortical bone) and the internal (trabecular) bone. This loss can appear as small pores to large cavities to the complete destruction of an entire bone. Abnormal bone gain takes the form of excess bone being deposited at various locations throughout the skeleton. These also vary in size from small exostoses (e.g., osteophytes) to large outgrowths.

DOI: 10.4324/9781003487944-8

Abnormal shapes involve deformity of bone such as bending of the long bones seen in rickets. The pathological conditions described below show some, or all, of these four bone changes and will be arranged into different 'types' based on their cause. These types are: congenital conditions, trauma, circulation disorders, joint diseases, infections, deficiency disorders, and tumors. This chapter will focus mainly on the more common conditions but will also include some rare disorders that are particularly interesting.

## CONGENITAL DISORDERS

Congenital disorders are pathological conditions that a person is born with. These conditions can be caused by genetic abnormalities, problems occurring within the womb, or both. This section will not discuss the causes of these types of conditions but will concentrate on their effects on bone and bone growth. One such condition is achondroplasia, a genetic condition that causes bones to be smaller than normal. This results in the skeleton of a person having a larger than normal head, fairly normal sized chest and pelvis (torso) but shortened arms and legs (limbs), particularly the upper arm bone (humerus) and upper leg bone (femur). In addition, teeth are often crowded, the forehead is prominent, the nose is flat, and the person has a maximum height of 4 feet. Several well-known (television and movie) actors had/have this condition including Michael Dunn and Peter Dinklage.

A congenital condition commonly seen in the sacrum is a gap between neural arches that is called spina bifida occulta. As described in Chapter 2, the sacrum is composed of (usually) five fused vertebrae (segments) that decrease in size from top (superior) to bottom (inferior). As with other vertebrae, those of the sacrum have a body and neural arch and, sometimes, the neural arches do not join in the center to form the median sacral crest, leaving a gap between the arch sides (see Figure 8.1A). This is a sign of a general condition called spina bifida, which in its most extreme examples affects the neural arches of some, or all, of the vertebrae. Since most people born with spina bifida died in childhood before modern surgery, those who survived in the past had a less extreme example that affected only the neural arches of the sacrum and probably did not suffer any harmful effects. For this reason, the

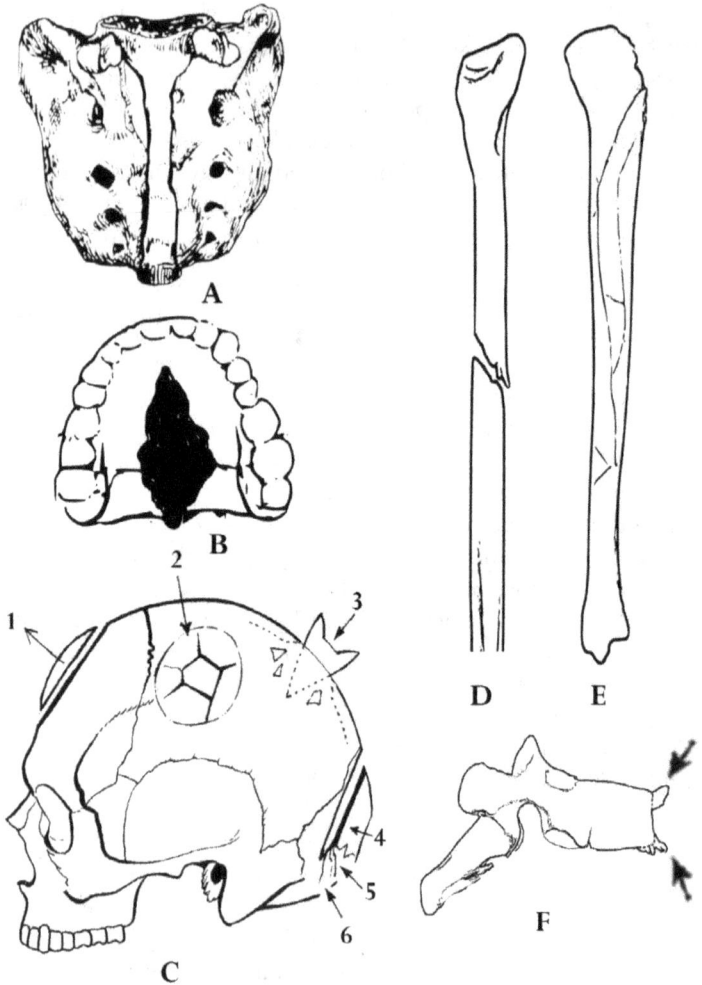

*Figure 8.1* Various Pathological Conditions Seen in Human Bone
Source: Images A and C from Figures 4.18E and 5.1A, respectively, of *Digging Up Bones* by DR Brothwell (1981) Ithaca, NY: Cornell University Press. Used with permission. See text for explanation.

condition in adult sacra with opened neural arches is called spina bifida occulta, with 'occulta' meaning hidden (symptomless).

Congenital dislocation of the hip is a condition where the hip socket (acetabulum) is more shallow than normal and the head of the femur is somewhat flattened. When walking, the hips have a tendency to dislocate causing (almost certainly) extreme pain and disability. People with this condition would not be able to perform many normal activities and would have to be cared for by other members of society.

Cleft palate is a congenital condition where the palatal part of the maxillae does not meet in the center (Figure 8.1B), causing any liquid or food in the mouth to be able to enter the nasal passage. Infants born with this disorder could drown on their mother's milk unless special care was taken during feeding. Although modern surgery can correct this condition, it is probable that most infants with this condition in the past died well before adulthood.

Sometimes the neural arches of the vertebrae, especially those from the lumbar region, do not fuse to the bodies, or become detached during adulthood (see Figure 7.2J). As described in Chapter 7, this is a disorder called spondylolysis and can be a congenital condition or be caused by a fracture due to accident or (more likely) strenuous activity. In modern people, this disorder usually does not have symptoms, but when it does the person can have a dull ach around the vertebra with the detached neural arch or have more severe symptoms like pain down one or both legs, muscle weakness, difficulty walking, and (more rarely) problems with bladder and bowel control.

Kyphosis of the spine is a condition in which the vertebral column curves forward to a degree greater than normal (see Figure 7.3G). Congenitally, this condition is caused by a hemi-vertebra (half of a vertebra) made up of the back (posterior) half of the vertebra in the spine. Similarly, scoliosis is a disease in which the vertebral column exhibits lateral curvature(s) and is S-shaped when viewed from the back (posteriorly). This can be congenital or due to a disease that destroys one side of a ver-tebra or several vertebrae. Congenitally, this condition can be caused by hemivertebrae located lateral to the sagittal plane in the spine. Both of these pathological conditions are rare in industrial populations.

Craniosynostosis is a condition where the suture lines between cranial bones fuse early before the brain has fully grown, resulting in a misshapen skull. A common form of this is early fusion of the sagittal suture causes the neurocranium to have a tent-like ridge along the midline. Sometimes the right or left half of the coronal suture fuse, causing the braincase to be flattened on that side while the other side has a more normal appearance. The same can happen to one side of the lambdoid suture, causing the back of the skull to be flattened on the affected side and normally shaped on the other side.

One last congenital condition is hydrocephaly, sometimes called 'water on the brain.' In this disorder, brain fluids (cerebrospinal fluid) build up deep within the brain, causing it to swell and put pressure on the walls of the braincase, which in turn causes the braincase to increase in size. Persons with this condition have an abnormally sized neurocranium but normal sized facial bones. If this disorder occurs in the womb, it is usually fatal to both the mother and the infant due to difficulty in giving birth. If the infant survives birth, most will die without modern medical treatments. When seen in a skeleton, the appearance of the head can be quite striking.

## TRAUMA

There are many hazards that people face in life that can cause injuries, both to soft tissue and bones. Simple tripping while walking, getting into an automobile accident, or slipping and falling from a high place all can cause injuries that may include broken bones. More dangerous hazards include interpersonal violence due to crime or, even more dangerous, warfare. People from the past faced some of these same hazards and a few that are not common in modern people (e.g., feuds, animal attacks) such that trauma has been part of human experience throughout time. As with other pathological conditions, trauma can help with understanding the cause of death of a person. Diseases described in this chapter usually help with establishing a manner of death that would be considered a natural. Trauma is unique in that it can help establish other manners of death, specifically murder, suicide, or accident. (There is a fifth manner of death called 'unknown' when death does not fit any of the other four types; most skeletons analyzed fall into this category.) This makes it unique among other pathological conditions and therefore it has been studied more thoroughly than many bone diseases.

When enough force is applied to bone, a break (discontinuity) will occur. Depending on the amount of force, the break may be just a crack in the bone (fracture line) as in lines 2, 5, and 6 in Figure 8.1C, or a complete break where the bone is in two separate pieces (simple fracture) as in the top (proximal) third of the fibula seen in Figure 8.1D or the bone may be shattered into many pieces (comminuted fracture) as in the side view of the tibia pictured in Figure 8.1E. The force that causes the break can come from many different directions: pulling (tension), pushing (compression), twisting (torsion), or side (bending and shearing). Any one of these can cause complete fractures with or without fracture lines (cracks) in adults. Sometimes these forces can cause what are called 'green stick fractures,' which are incomplete fractures (called infractions) that can occur when bone is supple enough to bend under the force of impact but not break in two. This fracture resembles the breaking of a fresh piece of wood, where fibers of the wood stick out from the side opposite from the force. They are most commonly seen in subadults when trauma occurs to a long bone, especially the clavicle.

There are several breaks that are so common that they have special names. The 'parry fracture' of the ulna is caused when persons hold their arms up, bent at the elbow, as a form of self-defense (e.g., to ward off a blow), causing the bone to fracture inwardly. Usually, parry fractures occur at the far end (distal segment) of the ulna; however, some of these fractures occur near the elbow. This type of traumatic injury is common in many deaths by violence (homicide). The other specially named fracture is the Colles's fracture. This is a break in the far (distal) end of the radius that happens when people fall forward with their hands out as they hit the ground. The weight of their own body causes the radius to break against the hard ground. This injury is most common in accidents, such as a fall from a bicycle.

The complete and incomplete breaks described above are interesting but equally interesting are the cracks that can form with them. These usually start near the point of impact, where they can either spread outwardly (called radiating fracture lines) or surround the point of impact (called concentric fractures).

Radiating lines are the most common type of fracture lines; these disperse outward, like an irregular sunburst, from the area where the force hits the bone (see lines spreading out in Figure 8.1C, arrow '2'). Generally, their center of radiation indicates the point of impact of the causative force. The other type of fracture line forms concentric rings around the area of impact such they are sometimes called hoop fractures (see circular fracture lines in Figure 8.1C, arrow '2').

Traumatic fractures causing breaks and lines are what usually come to mind when thinking about bone trauma. However, several other types of fractures also occur in the human skeleton. Pathological fractures are breaks that occur in bones that are weakened by disease. Thus, skeletons affected by osteogenesis imperfecta (a disease that makes bone weak and brittle) usually exhibit pathological fractures. Other stress fractures are breaks caused from overuse; athletes can suffer from these types of injuries, especially in their heels and legs due to repeated strain on these bones. Finally, fatigue fractures occur in bones that are exposed to intermittent stress over a long period of time. The most common example of this injury is stress fractures of the vertebrae in older persons; this is the disorder that causes the dowager's hump, a condition where the spinal column is angled so prominently forward as to appear to form a hump on the back. One last effect of trauma is myositis ossificans traumatica. This is a disorder where bone forms in muscles, tendons, and ligaments as a result of some trauma to the area.

There are three basic types of fractures, defined by the force that causes them: blunt force, sharp force, and projectile trauma. The first refers to any injury caused by a force that hits a bone over a wide area. Bones with blunt traumatic injuries usually exhibit both complete breaks and fracture lines (see arrow 2 in Figure 8.1C). Typically, these forces cause at least simple fracture wounds. However, comminuted fractures also are common, particularly when the bone is shattered by the application of excessive force. Although many instruments wielded as clubs result in blunt force trauma, any hard surface can cause this type of injury. Thus, impact of the body on the ground after falling from a great height generally results in crushed and shattered bone, as does impact during automotive, train, and airplane accidents.

Sharp force trauma is caused by sharp tools such as knives, axes, or any instrument with a sharp edge or point. These cause punctures (due to stabbing action), slices (due to slashing action), or notches (due to chopping action), all of which cause discontinuities and occasionally fracture lines. Punctures appear as pointed holes in bone where the sharp tip of a knife or other such instrument strikes into a bone. Slices (cut marks) are not only caused by slashing but also by back-and-forth action of a sharp blade across bone. Scalping involves this latter action where the removal of part of the skin with hair of the cranial vault leaves cut marks on the bone. (Scalping has been practiced by various societies in prehistoric and historic times.) Chopping not only causes notches but can also cause sections of bone to be cut off (displacement) as seen in arrow 1 in Figure 8.1C. With intense study and experimentation, characteristics of wounds caused by sharp tools can help identify the instrument used to injure bone.

Projectiles, particularly high velocity bullets but also spears and arrows, have such distinct wounding characteristics that they deserve a separate category of trauma. The results of bullets and points on arrows or spears (see arrow 3 in Figure 8.1C) impacting bone generally are complete discontinuities with both displacement and fracture lines. The most common projectiles that cause this type of trauma are bullets from firearms; however, any object that travels through the air and impacts with enough energy (e.g., arrows, spears) can cause projectile trauma.

There are several types of traumas that do not fit easily into any of the three types just described. These include trauma caused by stationary pressure (the most common of which is strangulation), generalized dynamic pressures (i.e., explosions), sawing, and trauma due to chemicals and heat. Generally, only the first three result in complete discontinuities with displacement of bone; also, they usually do not have a large number of fracture lines. Traumas caused by heat can exhibit both discontinuities and fracture lines

One of the most important topics surrounding bone injury is its relationship to time of death. After a break occurs and the person survives, the injured bone responds in several different ways. First, the veins and arteries that usually burst will leak blood and form a pool over the damaged area, called a hematoma, that helps to stabilize the broken pieces, especially after it clots (coagulates).

Second, trauma also causes the periosteum to form new bone by producing connective fibers that span between the broken surfaces. This flexible tissue can invade the hematoma to bridge the gap formed by the discontinuity. These fibers form the framework for the third action, which is the development of a callus composed of fibrous bone. If the break is sufficiently immobilized, this initial callus develops within the connective fibers as calcium matrix is deposited. Although composed of true bone, it does not have the strength of ordinary bone because it is neither well organized nor dense. The last stage in healing involves the replacement of fibrous bone with lamellar bone that is much stronger due to its greater organization and denser structure.

Sometime early during the above-described process, remodeling of the fracture edges occurs. The bone near the break begins to resorb, rounding the otherwise sharp borders of the break. In addition to rounding, occasionally pores develop in the area of the break. This porosity contrasts sharply with the smooth surface of the rest of the bone. The calluses formed during the last two phases cover these two changes and become raised and somewhat irregular areas over and around discontinuities. Under normal circumstances, calluses begin to form by the sixth week after injury, where they can remain visible for years after complete healing. In some cases, with enough time, all traces of fracture and callus can be resorbed, leaving no indication of a previous break.

All of this helps researchers identify the three timings of trauma: antemortem (premortem), perimortem, and postmortem. Antemortem trauma refers to trauma that occurred before death such that there is partial or complete healing of the injury. Although it does not give information about the cause of death, it can provide clues about the hazards faced by people. For example, many adult Neanderthals skeletons show healed fractures of blunt and/or sharp force trauma, indicating that they faced many dangers during life. Perimortem trauma refers to injuries that occurred at or around the time of death. Because of its timing, it does not show the characteristics of healing described above but can offer clues as to cause of death and even manner of death (homicide, suicide, accident). Finally, postmortem damage refers to injury that occurs after death. When this happens soon after death and the bone is still fresh, it looks very much like perimortem trauma. However, when a bone is dry, it often breaks at right angles to their long axis (as opposed to the angled break in Figure 8.1D) and do not show comminuted fractures (as in Figure 8.1E).

## CIRCULATION DISORDERS

When the blood supply to bone gets reduced or cutoff, the bone can die causing anything from small holes to total collapse of a bone at a joint. This condition is called osteonecrosis (literally bone death) and takes many forms. The simplest result of bone death is a small hole in the bone surface. These appear to be due to an activity, injury or disease that reduces or completely stops blood flow to part of a bone. They are usually very small (around 1 mm across) but can become large over time. These appear most often in the bones of the foot (tarsals and metatarsals) and may be caused by squatting while grinding grain (e.g., corn, wheat), which could limit blood flow to the foot bones.

More striking examples cause a large part of a bone to die, such as what happens when there is a fracture between the head and neck of the femur or a break in the middle of the femoral neck. These can stop blood flow to the bone by severing the arteries that supply blood to the proximal femur. The result is aseptic (non-infection) necrosis, where the head collapses unless circulation is restored. Similarly, Legg-Calvé-Perthes disease is a disruption of the blood supply to the head of the femur, which collapses due to aseptic necrosis. It is usually caused by trauma and occurs most often in children between five and nine years of age. A last circulatory disorder is osteochondritis dissecans. This disorder is found most often (90% of cases) in the knee, where a triangular piece of bone and cartilage separates from one of the femoral condyles and floats inside the joint. The piece of cartilage may heal over time, but the bone fragment remains dead and there is a depression in the condyle where the bone fragment detached from. The condition is found more often in males than females and first appears in adolescence and young adulthood.

## JOINT DISEASES

Probably the second most common disorder of the (adult) human skeleton (after generalized bone disease described below) is osteoarthritis. This condition causes boney outgrowths, described in Chapter 7, to appear in various locations in the body, but especially around the joints and on the bodies of vertebrae. When the joints are affected, a rim of extra bone can surround the joint surface such as around the scapular shoulder joint (glenoid fossa) or the hip socket (acetabulum). On the bodies of vertebrae, spurs of bone, called osteophytes, grow out from the edges (see Figure 8.1F) and are believed to be due to deterioration that comes with age. In addition, all of the joints around the vertebral facets can show these spikes of bone. In the past, anyone over the age of 50 would have these lesions; however, due to the less stressful conditions of modern life, their development in all present-day persons is not as common.

Another disorder of the spine is a pit that occurs on the top or bottom of vertebral bodies called Schmorl's node (see Figure 8.2A). These depressions in the vertebrae are caused by pressure on the vertebral column that pushes the gelatinous material in the intervertebral disks into the trabecular bone of the vertebral bodies. This causes a dent in the otherwise flat body surface. Generally, this disorder does not cause problems and is not usually surgically corrected. Pressure on the spine due to age, infection, trauma, or heavy lifting causes this intrusion and the formation of these pits. Schmorl's nodes are most common in the lumbar or lower thoracic vertebrae and have been seen in over 70% of skeletons examined after death.

General wear-and-tear on joints can cause the cartilage that separates the moving bones to become thinner and disappear resulting in the bones becoming smooth and dense from rubbing against each other. This is called eburnation and can be very painful. As with other joint disorders, there are a number of conditions that can cause this, including infections, trauma, heavy lifting, and just plain old age. In industrialized countries, joint replacements can fix this disorder but in the past people simply had to live with the condition and its associated pain.

Abnormal fusion between bones of a joint can occur due to a condition called ankylosing spondylitis. In this disorder, the joints between the vertebral facets and between the vertebral bodies fuse usually starting from the lumbar vertebrae and moving upward. In addition, the ligament that joins the spinous processes of the vertebrae (the interspinous ligament) and the ligament that runs down the back of the spinous processes (the supraspinous ligament) ossify.

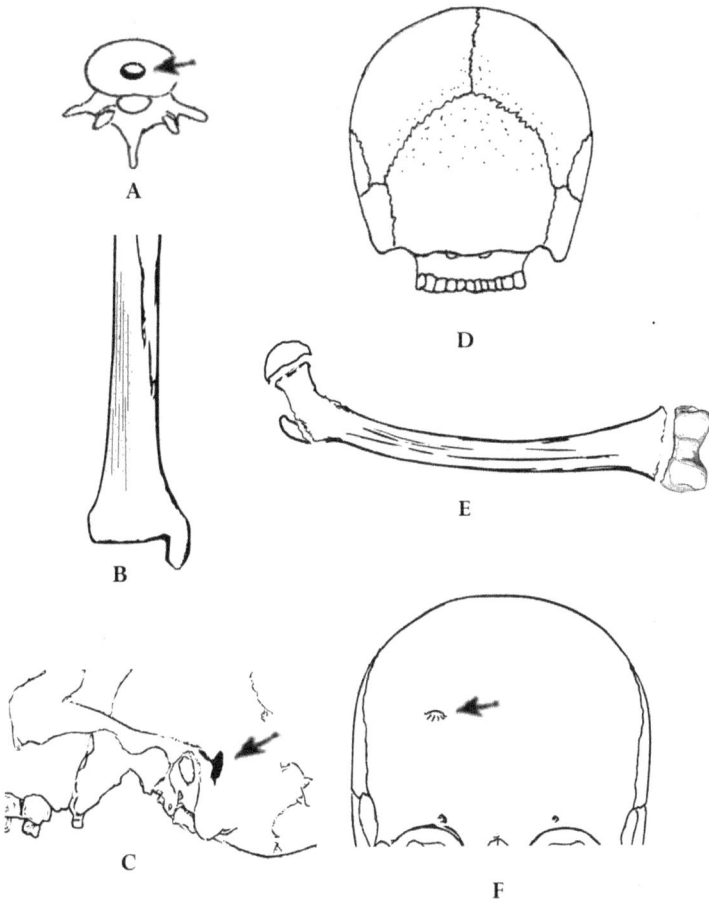

*Figure 8.2* More Pathological Conditions Seen in Human Bone (See Text for Description)

Other joints, such as the shoulder, hip, knee, and ankle can be affected but the hallmark of this condition is the fusion between the sacrum and the ilium (the sacroiliac joint).

One last joint disorder is diffuse idiopathic skeletal hyperostosis (DISH). In this condition, the ligament that runs down the front (anterior side) of the vertebral column (the anterior longitudinal ligament) ossifies, thereby reducing the flexibility of the spine. This ossification has a distinct appearance in that it looks a lot like flowing candlewax and is most common on the right side from the $7^{th}$ to $11^{th}$ thoracic vertebrae. In addition, other parts of the skeleton show ossification of the tendons and ligaments around joints as well as the cartilage of the neck and ribs. The earliest known example is a Neanderthal skeleton dated 40,000 to 73,000 years ago from what is now Iraq. The cause of this disorder is unknown, but it is seen in around 3% of modern skeletons but more than twice that amount in earlier populations.

## INFECTIONS

As most people know, infectious diseases are fairly common in everyday life. Caused by bacteria, viruses, and parasites, infected people experience a variety of symptoms and signs. Usually, people get over infections, with or without medicines, or die before the skeleton is affected. However, in some cases, bone changes do occur and do show what the individual suffered from during life. This section will first deal with bone infections that cannot be identified as a certain disease after which infections of a known type will be discussed.

The most common bone disease appears as small, long, boney mounds separated by thin lines with pores scattered between them that most often appear on the distal half of the tibia (Figure 8.2B) but can appear on any bone of the skeleton. The mounds are less than 1 mm above the outside (periosteal) surface of the bone, are usually less than 2 mm wide, and between 1 cm and 10 cm long (or longer). Occasionally, there are spatters of bone (plaques) scattered among the mounds and lines. These mounds are made up of lamellar bone that has been laid down by the periosteum reacting to an infectious organism or it can sometimes be caused by trauma. These changes are called periostitis when caused by an infectious organism but have also been called generalized bone disease since the causative organism is unknown. These lesions are surprisingly common in prehistoric and late historic skeletons but are also found on the bones of persons from industrialized countries such that they are probably the most common disorder of the (adult) human skeleton.

Infections can result in death of part of the bone, leaving an opening and occasionally a free-floating bone fragment. Figure 8.2C illustrates such a necrotic lesion behind the external acoustic meatus. Osteomyelitis is another bone disease where the microbe that causes it is unknown. In this infection, cortical bone both on the outer (periosteal) and inner (endosteal) surfaces are destroyed and covered by new bone (called an involucrum) that is larger, less organized and has a lumpier surface than the original bone. During the active part of the infection (before persons either die or heal on their own), pus is formed and vented to the outside, sometimes carrying bits of dead bone (called sequestrums). When the pus is generated inside the bone (the medullary cavity), a tunnel is formed thru the involucrum (called a cloacal opening) to allow the pus to exit the bone and into the surrounding soft tissue, where it may eventually push through the skin to ooze out of the body. Examples of osteomyelitis in prehistoric and historic skeletons can be quite striking in their size and appearance.

Venereal syphilis, Bejel (also called endemic syphilis), and yaws are three infectious diseases that are generally referred to as treponemal disease since they are all caused by variants of the *Treponema* bacteria. Bone changes usually only occur in the later stages of the diseases but have similar appearances. The most typical are pits in the bone surface (cortical bone) with lines that radiate outward in a star-like pattern such that they are called stellate scars. These occur most often in bones near the surface of the skin, especially the tibia but also the ulna, and cranial vault. In Yaws and Bejel, a common bone change is a forward (anterior) curving of the tibia, called a 'boomerang leg' in yaws and saber shin in Bejel. The skulls of persons with yaws show roughly circular pits that are caused by the destruction of cortical bone with reactive bone forming around their borders; this gives the lesions the appearance of a crater (e.g., meteorite crater). Venereal syphilis shows several bone changes in the later stages of the disease. The cranial vault and the area around the nose are common sites of destructive lesions with reactive bone formation. In the vault, these circular and noncircular lesions show the loss of cortical bone that is surrounded by reactive bone. In extreme cases, almost the entire vault has a lumpy appearance of mounds of reactive bone and cavities of active infection. In the nasal area, destruction of the nasal bones, anterior maxillae and palatal parts of the maxillae occurs, similar to that from leprosy.

Tuberculosis is an infection that most people associate with lung issues. Although the lungs are often affected, other body parts, including the skeleton, also show the disease. The hip and knee are common skeletal areas of infection, but the most common site is the vertebral column, especially the lower (inferior) thoracic and upper (superior) lumbar vertebrae. The disease destroys bone in a similar manner to osteomyelitis, but with less repair. Usually, the body of the vertebrae is 'eaten' away, causing the vertebral column to angle forward (kyphosis as in Figure 7.3G) and/or sideways (scoliosis). If the hip or knee is infected, both the acetabulum and femoral head of the hip, and condyles of both the femur and tibia show destruction. Reactive bone does form around the areas of infection, causing the joints and vertebrae to appear deformed and unsightly. One other place where reactive bone can occur is the inner surfaces of the ribs due to infection from the lungs causing the periosteum to form new bone of a honeycomb appearance.

Bone changes from leprosy can be very common although they occur later in the disease progression. The bones of the face, hands and feet are the most common sites where the *Mycobacterium leprae* or *Mycobacterium lepromatosis* bacteria attack. In the face, the bones in the nasal area are most affected with destruction and resorption of the nasal bones, nasal conchae, nasal septum (made up of the vomer toward the back and cartilage toward the front) and even the hard palate (palatal part of the maxillae). When untreated, the resorption will progress to widening the nasal opening (aperture) and destroying the inferior nasal spine and front (anterior) of the maxilla. In the hands, bone destruction starts with the tips of the fingers (distal phalanges) and progresses up the fingers but only rarely affects the bones of the palm (metacarpals). In the feet, destruction of the bone starts with the tips of the toes and progresses toward the ankle with the metatarsals resorbed and the joints of the ankle (tarsals) affected.

## DEFICIENCY DISORDERS

People of the past and in many third world countries face the problem of getting enough food to live full and healthy lives. Lack of food or lack of proper nutrients in food has led to problems with the physical and chemical processes needed to keep a person healthy, capable of having children, and even reaching normal height (stature) during growth. This section will deal with three common conditions: stunted growth, iron deficiency (anemia), and Vitamin D deficiency.

Stunted growth can be caused by not enough food during childhood and adolescence or poor-quality foods that causes growth to slow down. Studies of people in countries where there is both adequate and inadequate amounts of food for children show that those who did not have enough food in their subadult years did not reach the same height as those children with adequate amounts food. Even if enough food becomes available in early adulthood, people will not grow to normal size because the epiphyseal plates will have fused, and bone length cannot increase. This 'stunting' can be fairly small with the average (mean) stature of a nutritionally stressed group being 2% smaller than persons from the same population who had adequate food. On the other hand, an almost 11% difference in average (mean) height was seen when food was even more scarce.

One of the most common deficiency diseases found in prehistoric and late historic skeletons is anemia, particularly iron deficiency anemia. This is a condition where persons do not produce enough red blood cells because they do not have enough iron in their diet. This lack of iron causes red blood cells to become distorted from their original round shape, which causes them to break as they flow through capillaries and other blood vessels. Once broken, these cells must be eliminated from the body and replaced. The normal response to both low levels of iron and faster loss of cells is to increase the production of red blood cells. This causes the diploe of the skull where most red blood cells are made, to expand to meet the greater demand. This expansion can result in the pushing of the diploe against the inner and outer tables of the bones of the cranial vault and occasionally the face. The result of this pushing is the formation of pores and (sometimes) outgrowth of the diploe.

In dried bone, this condition (termed porotic hyperostosis) usually is found on the occipital (see Figure 8.2D) but can be located on the parietals (particularly near the sagittal suture), frontal (near bregma), or any other bone of the vault. In the face, the upper (superior) eye orbits often show these pores and occasionally diploe outgrowths. The pores come in varying sizes, from small pinpricks to larger openings that can be as much as 3 mm in diameter. Although found more often in prehistoric than contemporary people, this condition may also appear in persons suffering from general malnutrition, a possibility among people from the lower socioeconomic classes. It can also be due to an inherited anemia, such as sickle cell anemia, seen most often in Blacks.

A last deficiency disorder is caused by not having enough Vitamin D and/or not enough calcium. Vitamin D aids in the absorption of calcium from the diet, which is used in many parts of the body including the bones of the skeleton. Calcium is one of the major minerals that, along with phosphorus and magnesium, make up bone. Deficiencies of either of these two can cause the bones to become soft such that they distort due to weight during walking or due to muscle contraction (e.g., the muscles on both sides of the pelvis constantly pull against each other to keep it level). Soft bones cause rickets in infants and children (subadults) and osteomalacia in adults. In subadults, walking puts weight on the lower limbs and causes them to become bowed (see Figure 8.2E) and their metaphyses to flare and become wider than normal. Other bones are also affected. In adults, the weight bearing bones do not usually bow but the pelvis can become distorted causing problems with walking and (in women) childbearing. Although too little vitamin D is rare in industrialized countries, persons from the third world countries are at risk for this deficiency due to a general lack of food.

## TUMORS

When most people think of tumors, the automatic response is cancer. Tumors are caused by abnormal growths of cells that can occur anywhere in the body, including the skeleton. Tumors can be malignant, spreading to different parts of the body (metastasizing), and often resulting in the death of the individual. However, some tumors are benign in that they do not spread and cause no harm to the person with the disorder. A common nonmalignant tumor is a button osteoma (button lesion) that usually occurs on the braincase of the skull and is visible on live persons particularly if located on the frontal (see Figure 8.2F). Malignant tumors of the bone, usually called osteosarcoma, are rare in the human skeleton but can be very dramatic.

# DENTAL PATHOLOGIES

The teeth, like the bones, are also prone to pathological conditions. The most common is the cavity (caries) caused by decay of the enamel and often the underlying dentin. These lesions occur anywhere on the teeth, from between the teeth (1 of Box A in Figure 8.3), to on the neck (2 of Box A in Figure 8.3), to above the neck (3 of Box A in Figure 8.3) to on the chewing (occlusal) surface (4 of Box A in Figure 8.3). If left untreated, the tooth will continue decay to the point that the entire crown is destroyed, leaving only the tops of the roots jutting from the bone of the jaws (incisor in 1 of Box D in Figure 8.3).

Another common dental pathological condition to consider is tartar, also called calculus (Box B in Figure 8.3). This is a hardened form of dental plaque, which is a film made up of proteins, bacteria, and other substances. As plaque builds up, it absorbs minerals (mainly calcium phosphate) from saliva and a fluid released by gums inflamed by the proximity of plaque (gingival crevicular fluid). Although this process kills the bacteria in the plaque, it forms a good base for the formation of more plaque which can harden into more calculus and cycle repeats itself over and over. As can be seen from Figure 8.3, calculus can build up from a small amount (1 of Box B in Figure 8.3), thicken to a considerable degree (2 of Box B in Figure 8.3) and eventually cover the entire tooth (3 of Box B in Figure 8.3). As the calculus thickens, it pushes against the gums causing periodontitis, a condition where the gums become inflamed and diseased followed by destruction of that part of the maxilla and mandible that hold the teeth (called the alveolar bone).

In the absence of modern dental care, the destruction of the bone around the base of the teeth causes the bone to recede, exposing first the neck and then the roots of teeth. This is called alveolar resorption or periodontal disease. Box C of Figure 8.3 shows the stages of this process, starting with none (1 of Box C), to exposure of the neck (2 of Box C), to exposure of the tops of the roots (3 of Box C), to exposure of much of the roots (4 of Box C). When resorption is severe, teeth may fall out (exfoliate), leaving a gap where the tooth once was.

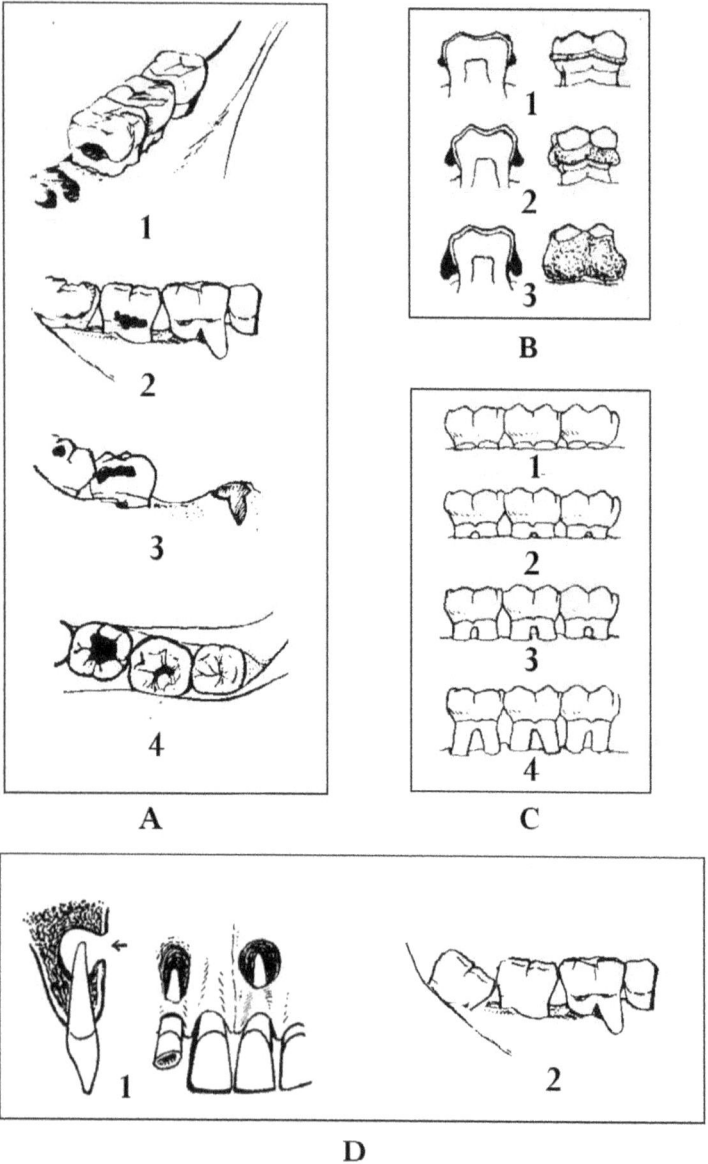

*Figure 8.3* Various Pathological Conditions Seen in Human Teeth (See Text
    for Description)
Source: Taken from Figures 6.12, 6.14, and 5.15A of *Digging Up Bones* by DR
Brothwell (1981) Ithaca, NY: Cornell University Press. Used with permission.

Another common pathology of the teeth are abscesses (1 of Box D in Figure 8.3). When decay reaches the point that the pulp cavity is exposed, bacteria can get into the bone of the jaws by traveling down the tooth roots until it reaches the tip. Here it begins to multiply while the body tries to kill it using its natural defenses. Over time, a pocket of pus forms at the tip of the root, called a periapical abscess, that continues to grow while the bacteria destroy the surrounding bone. Eventually, when enough of the bone is destroyed, the pus pushes through to the surface (1 of Box D in Figure 8.3), either inside or outside of the mouth to drain. In the upper or lower jaws of persons suffering from an abscess, the vent is usually easily seen and sometime even the cavity where the infection was located.

Impaction is a dental condition where a tooth grows crookedly and does not erupt vertically out of its crypt into the mouth. Instead, it can angle into another tooth (2 of Box D in Figure 8.3), come out of the side of the crypt, or emerge through the anterior or posterior bone of the jaws. On occasion it can grow out of the crypt in the opposite direction from normal and try to emerge through the bottom of the mandible or into the nasal cavity of the maxilla. Generally, these are very painful and, in modern societies, are removed.

Another type of pathology is dental hypoplasia. These are either pits in the enamel of the teeth or lines of thin enamel running across the smooth tooth surface, parallel to the chewing surface. This thinning of the enamel is due to a disturbance during that part of growth where enamel is being formed on the teeth in their respective crypts. They can occur on any tooth but are most often seen on the outer (labial) surface of the incisors and canines. It is not unusual for one individual to have more than one hypoplasia on each tooth. These pathologies are unique in that the age of the person when the growth disturbance can be estimated. For example, looking at Figure 5.1 of Chapter 5, if a line appears half-way between the tip (cusp) of the permanent upper canine and its neck, an age of 4 years is indicated because that is when the top half of the enamel is formed. Information about when growth disturbances occurred is another example of how the life of a person can be interpreted from their remains.

## SUMMARY

1 The human skeleton sometimes shows changes due to diseases and other pathological conditions.

2 There are four basic types of changes: abnormal bone loss, abnormal bone size, abnormal bone shape and trauma.

3 Congenital conditions are those that are seen at birth and are caused by genes or problems during growth in the womb or both.

4 Trauma is injury to bone caused by an outside force such as projectiles (bullets, arrows), sharp instruments (knives, axes) and blunt objects (clubs, hard ground in a fall).

5 Circulation disorders are pathological conditions that cause bone death (necrosis) and can affect any bone in the body.

6 Joint diseases take the form of osteophytes around vertebral bodies and facets, ossification of ligaments, and loss of bone that is part of a joint (Schmorl's nodes in vertebral bodies).

7 Infections are caused by any number of pathogens (bacteria, viruses, parasites) and can cause bone destruction with, or without, reactive bone growth.

8 Deficiency disorders are those caused by insufficient foods, the most common of which is iron deficiency anemia.

9 Tumors are examples of abnormal bone gain and one of the most common of which is the button osteoma.

10 Teeth are also affected by pathogens that can cause cavities (caries) that cause abscesses and plaque.

11 Plaque can push against the gums causing the underlying bone to resorb (periodontal disease).

12 Impaction of teeth occurs when a tooth grows crookedly and pushes against another tooth or the bone of the jaws.

## FURTHER READING

The best books for diseases of the bone are: JE Buikstra, ed. *Ortner's Identification of Pathological Conditions in Human Skeletal Remains* (Amsterdam: Elsevier, 2019). AC Aufderheide and C Rodriguez-Martin, *The Cambridge Encyclopedia of Human Paleopathology* (Cambridge, UK: Cambridge University Press, 1998). A less intimidating volume is: C Roberts and K Manchester, *The Archaeology of Disease*, 2nd Edition (Ithaca, NY: Cornell University Press, 1997).

# CULTURAL MODIFICATIONS

As seen in previous chapters, there are a large number of differences in the human skeleton caused by different biological forces: genes, nutrition, stress, diseases, and others. This chapter will discuss variations in bones and teeth caused not by biology, but by humans. Many societies around the world intentionally (and sometimes unintentionally) modify the human body to one degree or another. In the soft tissue, these include tattoos (ink inserted into the skin), penile implants (insertion of pebbles into slits in the head of the penis), clitoridectomy (partial or total removal of the clitoris), piercings (ears, nose, lips), elongations (objects inserted into the fleshy parts of ears and lips), and others. In addition to soft tissues, bones and teeth are also modified, sometimes intentionally and other times as a byproduct of a societal practice.

As with soft tissue modifications, there are a number of modifications to bone and teeth practiced by various societies in various locations throughout time. One of the oldest (perhaps THE oldest) and most widespread is the modification of the shape of the braincase. Cranial modification takes many forms from flattening parts of the braincase to causing the cranial vault to elongate backward (posteriorly) and upward (superiorly). Less common are dental changes where the teeth are sharpened, chipped, grooved, or even ritualistically removed. Surgery is also seen in many societies, such as the removal of sections of the skull vault (called trephination). In addition, there are less common and less widespread modifications such as removal of parts of the body (i.e., amputations) and foot binding (seen only in the orient). Some of these modifications do not appear to

DOI: 10.4324/9781003487944-9

cause pain to the affected individuals while others are well known to be painful since they were performed before modern painkillers (anesthetics) were available. This chapter will describe the most common bone and tooth modifications caused by cultural practices seen in many parts of the world, and a few of the less common modifications. Only those seen in prehistoric and premodern societies will be discussed since changes due to modern surgery and dentistry are well known to readers.

## COMMON MODIFICATIONS

Artificial cranial modification is the most common change made to bone by humans. Changing the natural shape of the human skull has been found in the prehistoric populations of many areas of the world, including North America, South America, Europe, the Middle East, Asia, Oceania, Africa, and even Australia. The first recorded example of this practice is in the Neanderthals from Shanidar cave in what is now Iraq. Authorities agree that deformation occurs when pressure is exerted against the skull, usually starting with infants. This pressure can be circular (called circumferential) or flat as with a hard cradle board.

Circumferential pressure is applied by wrapping bands around the head, with or without pads or tablets (see Figures 9.1B4 and 9.1A). This results in a molding of the braincase upwardly with, or without, a circular depression. The second and more common source of pressure involves applying flat surfaces (tablets) to heads using (at least) three different methods. First, two "free" tablets, which can be hard and flat or flexible, are tied together with one against the forehead and the other against the back of the skull (see Figure 9.1B1). Second, one tablet can be "fixed" (as in the use of a cradleboard) and another tablet tied over the frontal bone (see Figure 9.1B2). Skulls subjected to pressure from the front as well as the back show flattening of both the frontal and occipital bones (see Figure 9.1C). Last, while an infant's body is firmly held against a fixed tablet such as a cradleboard, the forehead is pulled against the tablet using a band or a band with pads (see Figure 9.1B3) causing the back, but not necessarily the front, of the skull to flatten.

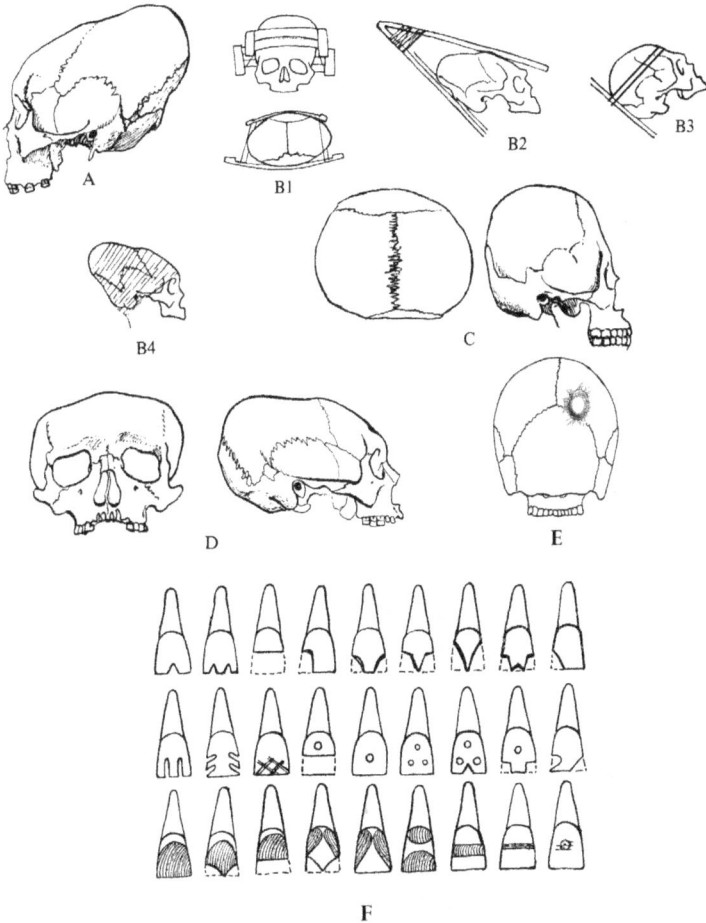

*Figure 9.1* Different Methods and Results of Artificial Cranial Deformation and Tooth Modification (See Text for Description)

Source: Except for E, all images are from Figures 2.27 and 4.22, respectively, of *Digging Up Bones* by DR Brothwell (1981) Ithaca, NY: Cornell University Press. Used with permission.

The degree to which these types of modifications are intentional or unintentional is unknown. Some researchers feel that flattening of only the rear of the skull was unintentional, especially when the deformation is minor. On the other hand, most skeletal biologists feel that deformation involving the use of pressure on both the frontal and occipital, or circumferential pressure almost certainly was done to intentionally change the shape of the skull.

The reason for this cultural practice is considered to be mainly cosmetic but other reasons are given by various societies. Some cultures feel that this practice increases the vitality of those so deformed. Others use skull modifications as a tribal marker or to indicate high social status and/or simply distinction and superiority. Even religious significance and fierce appearance in battle have been given as reasons. It has even been suggested that the abnormal shape of a ruler's head might have led to the development of this practice in other members of a society who wished their children to imitate the group's leader(s).

Experts are almost unanimous that cranial deformation affects neither final brain size nor function. Additionally, some authors feel that this practice has no effect on other cranial elements, such as the dentition and face. However, the extreme deformations seen in some specimens (see Figure 9.1D) makes it likely that such individuals may have suffered from behavioral or neurological problems. Similarly, some skeletal biologists have noted complications such as necrosis (death) of parts of the occipital bone, and premature closure (synostosis) of the cranial sutures. Other changes include modification to the shape and measurements of the face, arrangement of blood vessels inside of braincase, lambdoid sutural complexity, and form of the cranial base.

Trephination (sometimes spelled trepanation), which is the next most common bone modification seen worldwide, is the removal of part of a braincase. The earliest known examples of this practice appear in Europe during the Neolithic period (7000–3000 BC), and it is described in the writings of the great Greek physician Hippocrates of Kos (460–377 BC). Many examples of this surgery have been found throughout time with some third-world societies still performing trephinations. Geographically, it is found in Australia, Europe, North and South America, the Middle East, Africa, China, and on a number of Pacific islands. Societies perform this surgery to treat a variety of conditions such as migraines, epilepsy, and trauma to the head or to release evil spirits which were causing chronic headaches or other conditions.

There are four methods used to remove the skull sections. In some cases, the bone is scraped away until the inner cortical surface is opened and the soft tissue covering the brain (dura mater) is exposed (see Figure 9.1E). Another involves drilling holes in a circle and cutting between the holes to remove a disk of bone, while some practitioners use a round tool to cut through the inner and outer tables and remove the resultant bone. In a last method, four lines are cut into the bone: two parallel to each other one to two inches apart and two more parallel lines at right angles to the original lines. Thus, the lines intersect at the corners of a rough rectangle and the rectangular section of bone is removed. In most societies, this procedure was done without painkillers although substances that function as a local anesthetic have been observed in some societies. In historically documented cases, the operation is said to cause little pain, except when the scalp is cut and peeled back from the area to be operated on. Frontals, parietals and occipitals are the bones most commonly trephined, with the temporals only occasionally operated on. (Uniquely, trephinations on the tibia and ulna are performed on the island of Uvea in the south seas.)

There are examples of this surgery in the archaeological record that show patients survived this procedure since the skull bones have evidence of healing (i.e., new cortical bone covers the diploe). There are also historic accounts of persons who survived the operation, even after having multiple trephinations. Despite this, there are many cases where the diploe is exposed with no evidence of healing, indicating that the patient died. Although there are few studies of the number of patients who died from the operation, it is usually reported that 50% to 90% of them recovered, with most estimates on the higher side of this range.

Dental modification is the last most common alteration performed by societies. This takes three different forms: those that change the shape of teeth, those that remove teeth, and those that correct a pathological condition such as tooth decay. There are a large number of shape changes that people make to their teeth and Figure 9.1F shows a select few of these. As can be seen, sharpening to a point or points, making the teeth appear to have saw-like edges and sides, holes drilled through them, and etching cross-hatched patterns on the outer (labial) surface are the most common changes done by societies. Less common changes involve the inlay of stones such as turquoise or malachite on the outer surface. By contrast, some societies choose to remove teeth (called evulsion or ablation) instead of, or in addition to, tooth changes. This is the intentional removal of healthy teeth, usually the upper incisors and canines. In some cases, the teeth are broken off and the roots remain in the jaws while in other cases both the crown and root are removed. (Prehistoric dentistry will be described in the next section.)

The practice of changing the shape of teeth is sporadically seen in China, Indonesia, Africa, and Middle and South America. In these places, usually only the shapes of incisors are modified but the shape of the canines also can be changed and even the molars (these are usually ground flat not pointed). In most societies only the upper teeth are modified, while in others both upper and lower teeth are changed. Tooth evulsion is seen in Africa, Asia, the Pacific Islands (Oceania), and even Scotland. Reasons given for both shape changes and tooth removal vary from spiritual (religious) reasons, tribal markers, beauty, and mourning the loss of a loved one. Although mostly seen in prehistoric and late historic societies, tooth modification is seen in some societies even today.

## RARE MODIFICATIONS

Amputations, the removal of parts of a body such as arms or legs, are rare but did occur even in early societies. These operations are sometimes done for ritual reasons but more often for medical reasons such as the removal of a diseased appendage due to frostbite, crippled fingers or toes, or limbs with complicated fractures. A common form of ancient cave art appears to show amputations of fingers in what are called hand stencils that have both regular length fingers and shortened fingers (see Figure 9.2A). Although many societies practice such removals, the actual number of times such amputations are carried out is fairly rare. The appearance of the bone remaining after an amputation is similar to that seen in trephination; that is, the trabecular bone of the medullary canal is exposed at the time of the operation. If the person survives, cortical bone will cover the cut surface, and if the patient lives a long time after the amputation, the remaining bone will become reduced in size (will atrophy). Figure 9.2B illustrates this on a Neanderthal male from Shanidar cave in Iraq, where the end of

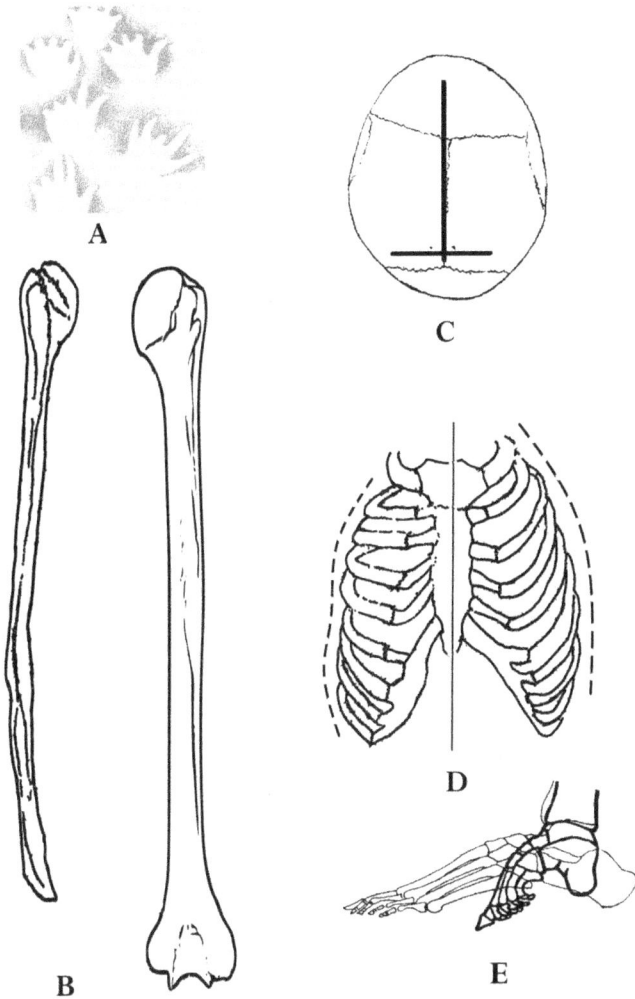

*Figure 9.2* Examples of Other Cultural Modifications (See Text for Description)

Source: Image E from Figure 2.27 of *Digging Up Bones* by DR Brothwell (1981) Ithaca, NY: Cornell University Press. Used with permission.

the humerus is rounded, and the entire bone is thinner and more delicate (reader's left) than a normal humerus (reader's right).

Dentistry has been practiced by various societies throughout time. The most common result of early dentistry are grooves that appear between the teeth due to the use of what are now referred to as toothpicks. These have been seen in Neanderthals as well as prehistoric and historic populations from Africa, Asia, Australia, North America, and the Pacific Islands. Although it cannot be known why prehistoric humans used these, the most likely reason is that toothpicks helped to dislodge trapped food and/or relieve pain due to inflamed gums.

Although less common, another example of early dentistry is tooth drilling presumably to remove cavities (caries) and/or to relieve dental pain. Currently, the oldest known example of tooth drilling in what appears to be an attempt to remove decay dates 14,160 to 13,820 years ago in northern Italy but drilled teeth have been found in prehistoric societies of what is now Pakistan (9,000–7,500 years ago), the American Southwest (around AD 1025), Alaska (AD 1,300–1,700) and Slovenia (6,500 years ago). A deeper look at these examples shows that not all of the drilled teeth from Pakistan had cavities so the reason for the drilling is unknown at this site. The drill holes in both of the teeth from the southwest and Alaska went into the root canal and both teeth had infections in the bone around the root tips (see periapical abscesses discussed in Chapter 8). This indicates that the drilling may have been done to allow the pus from the abscess to vent through the root canals to the outside, relieving pain caused by the pressure of the contained pus. The drilled tooth from Slovenia had a filling made from beeswax while the specimen from Italy shows that the drilling was done to remove a cavity (caries). These drillings were done in societies without metal (e.g., copper, bronze, iron) meaning that the most likely tool was a stone drill of some sort.

T-Sincipital Cautery is a method of burning (cauterizing) the scalp on the superior part (top) of the skull with the intention of relieving epilepsy or other nervous conditions (e.g., convulsions). The practice involves cutting the scalp in either a T-shape or cross and pouring hot oil into the wound; the heat of the liquid both stops any bleeding and sterilizes the wound, thereby reducing the chances of infection. The injured underlying bone shows grooves (long depressions) in the original shape of the incision (see Figure 9.2C). First identified in Europe, it has been found in Central Asia and Africa, and at least one example has been identified from South America (Peru).

Torso Modifications occurred because of tightly laced corsets that were worn in Europe between the 1500s and 1800s. These garments were mainly fashion accessories that reduced the size of the waist in women but were also used by both sexes as a brace to stabilize a diseased vertebral column. Several bone changes are seen in the people who wore them. The spinous processes of the thoracic vertebrae are seen to deviate to the side (laterally), and the ribs are less C-shaped. This last change results in a slightly concave rib cage (Figure 9.2D: rib cage on reader's left; dashed line emphasizes concavity) instead of the convexity that is normal (Figure 9.2D: rib cage on reader's right; dashed line emphasizes convexity); also, this causes the width and depth of the rib cage to be almost equal, not flattened front to back (anterior-posteriorly) as is normal. In addition, ribs also show fractures due to the pressure of tightening the corset, apparently helping to make the ribs straighter. When used for medical reasons, pathologies of the vertebral column (e.g., destruction of vertebral bodies due to tuberculosis) are also seen in conjunction with the corset.

Another practice that modifies bone shape is neck elongation in women. Although seen in Ndebele women of South Africa, bone changes have only been observed in the Kayan and Padaung women of Myanmar (Burma). Starting in youth, a brass coil is wound around the neck from the top of the shoulders to the bottom of the chin. As time goes on, longer coils are added until the woman's neck appears to be twice as long as normal. This causes the shoulders to be forced downward to the level of the second thoracic vertebra instead of the seventh cervical vertebra and the collar bone (clavicle) to be deformed. In addition, the distance from the lower border of the nose opening to the lower border of the mandible is smaller on the average from women without neck coils. Also, the incisors are more slanted forward (anteriorly) causing buck teeth (overjet) in women with neck coils.

A final cultural modification that deserves description is foot binding found only in the orient, especially China (but also in parts of Korea and Japan). This is a practice where the feet of female children are tightly bound to prevent normal growth. The reasons given for this practice vary from it being a sign of beauty to making it possible for girls to only do handwork such as basketry and embroidery (and not work requiring standing and walking), and even a rite of passage. First seen in the 10$^{th}$-century of China, it was originally limited to high status girls/women but eventually was practiced on all women except for those destined to work in agricultural fields (presumably so their small feet did not interfere with planting and harvesting). The practice persisted until it was criticized by European missionaries, after which it slowly declined until it was made illegal in China by the 1950s.

The process of foot binding commenced when girls were any-where from four to nine years old. It involved bending the toes, except the big toe, downward and backward so that they were pressed into the sole of the foot until they (the phalanges) broke. The foot was then forced to be in line with the leg until the arch broke (the metatarsals), after which it was bound by a cloth keeping the toes and foot in this position. The binding was renewed once a day or several times a week until the foot retained the form without binding. Figure 9.2E shows the effect of this practice on the foot bones (superimposed over a normal foot), especially the metatarsals and phalanges. Notice the foot is arranged in such a position that the girls/women walked on the front part of the foot without the heal (calcaneus) ever touching the ground.

## SUMMARY

1   Body modifications are common in societies worldwide and can be made to soft tissue as well as bones and teeth.

2   Cranial deformation is probably the most common bone change seen worldwide and throughout human history.

3   Trephination, the removal of part of the braincase, is a common form of surgery practiced throughout the world and throughout time.

4   Modifying the shape of teeth, most commonly the incisors, is also seen in different parts of the world and in prehistoric and historic peoples.

5   Less common bone modifications are: amputations of bones or parts of bones, tooth pick grooves and dental drilling, burning

(cauterizing) the scalp causing a T-shaped scar on the top of the skull, torso modifications due to corsets, facial and dental changes due to neck coils, and foot binding.

## FURTHER READING

For an overview of most cultural modifications: R Redfern and CA Roberts. Trauma. In: JE Buikstra, ed., *Ortner's Identification of Pathological Conditions in Human Skeletal Remains* (London, UK: Academic Press, 2019), pp. 264–273. For cranial modification: V Tiesler, *The Bioarchaeology of Artificial Cranial Modifications: New Approaches to Head Shaping and Its Meanings in Pre-Columbian Mesoamerica and Beyond* (Berlin, NY: Springer, 2013). For trephination: C Gross, *A Hole in the Head: A History of Trepanation* (https://thereader.mitpress. mit.edu/hole-in-the-head-trepanation/, 2012). For dental modification: S Bernett and J Irish, eds., *A World View of Bioculturally Modified Teeth* (Gainesville, FL: University Press of Florida, 2017). For amputations: EH Ackerknecht, Primitive Surgery in D Brothwell and AT Sandison, *Diseases in Antiquity* (Springfield IL, Charles C. Thomas, 1967). For early dentistry: G Oxila and others, Earliest evidence of dental caries manipulation in the Late Upper Paleolithic (*Nature - Scientific Reports*, 5: 121505, 2015). For T-Sincipital Cautery: RL Moodie, A Variant of the Sincipital T in Peru (*American Journal of Physical Anthropology* 4(2):219–222, 1921). For torso modifications due to corsets: R Gibson, Effects of Long Term Corseting on the Female Skeleton: A Preliminary Morphological Examination (*Nexus, The Canadian Student Journal of Anthropology* 23(2):45–60, 2015). For neck coils: D Chawanaputorn and others, Facial and dental characteristics of Padaung women (long-neck Karen) wearing brass neck coils in Mae Hong Son Province, Thailand (*American Journal of Orthodontics and Dentofacial Orthopedics* 131(5):639–645, 2006). For foot binding: H Seymour Levy, *Chinese Footbinding: The History of a Curious Erotic Custom* (Gravesend, UK: Bell Publishing, 1967).

# 10

# SPECIALTY METHODS

As discussed in Chapter 1, the answers to the various questions asked about persons represented only by their skeletons can have different uses depending on what a person wants to know about the dead. The chapters up to this point have presented basic methods for answering the "big four" and the rest of the eight questions listed in Chapter 1. However, there are a number of other methods beyond basics that answer other questions about skeletal populations. These methods are special to one or more of four subareas of interest to human skeletal biologists: forensic investigations, lifeways of past populations, diseases in the past, and the evolution of prehumans to modern humans. This chapter provides a brief overview of specialty methods used to study skeletons in these four areas that have not been described up to this point. Interested readers can review the books under Further Reading for more information on these subjects and their special methods.

## FORENSIC INVESTIGATIONS

Human osteology can contribute to forensic investigations when possible victims of crime or other violence are represented only by their skeletons. In such investigations, the answers to the 'big' four (ancestry, sex, age-at-death, stature) are used to get a possible identity of a person from a missing persons file. Other skeletal analyses can help with a positive identification (i.e., the name of the person represented by the skeleton) and, if trauma is present, perhaps information on cause and manner of death. This helps to

DOI: 10.4324/9781003487944-10

bring justice to victims of murder, genocide, and terror attacks as well as help identify persons killed by natural catastrophes (e.g., hurricanes, tsunamis, floods) or human disasters (e.g., airplane crashes).

One of the first tasks in this type of investigation is to determine if human bones are medicolegally significant; that is, did the person represented by a skeleton die within the last 50 years (approximately) while not under the care of a doctor. Persons who fit this definition may be victims of a crime and their deaths should be investigated in hopes of finding the person(s) responsible. (Fifty years is used since it is the approximate length of adult lifespan of a possible offender.) There are around seven traits that bones can have that indicate medicolegal significance: color, texture, hydration, weight, condition, fragility, and soft tissue. Generally, medicolegally significant bones are the same color of fresh bone which is yellow-white or white, their surface texture is smooth not pitted like bone that has been exposed to the outside or buried, they still have fats that saturate bones when in the body, they are heavy because of these fats (not light in weight where the fats have evaporated or leeched out while buried or exposed to the weather), they are in good condition and not fragile like old and buried bone, and (lastly) they have soft tissue attached such as cartilage, tendons, ligaments, and others (skin, internal organs). In addition, medicolegally remains usually are not found with prehistoric artifacts such as pottery or potsherds (broken pieces of pot), are not found in formal cemeteries, are not embalmed, do not have the bone modifications described in Chapter 9, but may have modern medical appliances such as hip replacements, gold crowns, and modern false teeth. These traits can be used to decide whether the skeleton needs further investigation or is too old to be of interest to the legal community.

Once remains have been determined to be medicolegally significant, an estimate of how much time has passed since the death of the individual (called postmortem interval, or PMI) should be made; that is, how long have the persons remains laid undiscovered. This interval, along with the 'big four', can help identify a person on the missing persons' file. There are a number of methods used to estimate this interval but most rely on the amount of time needed for bodies to reach different amounts of soft tissue loss. This amount of time has been standardized by studying many cases where the last date of persons being seen and the date of the discovery of their remains are known. These studies have shown that temperature, humidity, and accessibility to animals (insects, dogs, coyotes, and others) are the major factors that affect the speed at which soft tissue disappears and (eventually) bone deteriorates. Studies have shown that bodies laying outside in high temperature and humidity decay faster than outside bodies in colder and drier climates. For example, bodies lying out-of-doors in humid Tennessee would have soft tissue removed so that bones are visible within about a month while that amount of decay would take 2.5 to 4 months in the arid weather of Arizona. Estimates of PMI can eliminate some missing persons because the state of decay is either too great or too little for the time that the person has been missing.

A problem not usually encountered by most skeletal biologists, but is often part of a forensic investigation, is the removal of soft tissue from human remains. Many forensic cases involve bodies that have some, or much, soft tissue such as (usually dried) skin and internal organs as well as muscles, tendons, ligaments, and cartilage. These must be removed before the skeleton is analyzed since they can hide some of the features needed to answer the 'big four' as well as indicators of cause and manner of death. There are many methods for removing these tissues. The most common is cutting away as much of the tissue that can be done without damaging the bone after which it is soaked in water for several days or weeks whereupon the remaining tissues become so soft that they are easily rubbed off. Next, the bones are 'cooked' in hot water so that any remaining soft tissue can be easily removed and the natural oils and fats in the bones boil out. There are various ways of speeding up this process, but the end result is the same: dried bone whose surface is easily examined.

Another subject that arises in a forensic investigation is modifications done to the skeleton after death (postmortem). Understanding the forces that modified a dead body (and therefore skeleton) is the subject of the field of study called taphonomy. There are two different forces that can affect a body after death: human and natural. The changes caused by humans are usually done by murderers to the bodies of their victims to show their contempt for them and/or hide their identity. Dismemberments (removing the head, arms, and legs) using cutting instruments, especially saws, is one of the main human-caused taphonomic changes. Encountering such bodies has led to research into the effect of saws on bone such that things like direction of cut (anterior to posterior thru the legs), number of teeth per inch on the saw, saw blade width, blade type (crosscut vs. rip), blade shape (straight or circular), and source of energy (hand or electrical power) can be told.

Similarly, nature causes postmortem changes to bones due to many different agents. Common and important agents are the various animals, especially meat eaters like dogs and coyotes, who scavenge dead bodies. Their activities can cause puncture marks into bones, shallow pits, and scoring (scratches) from teeth that might be mistaken for trauma marks described in Chapter 8. The effect of other scavengers, especially racoons and skunks and rodents (such as rats and squirrels), have also been studied so that any marks they leave on bone will not be misinterpreted. Damage due to fire, weathering, burial, and water transport also have been studied so that as much information about the deceased can be known.

In a forensic investigation, the importance of bone trauma is greater than in other research since it may provide information on cause and manner of death (a main goal of any forensic investigation). As discussed in Chapter 8, the timing of injury is either antemortem (before death, usually with healing so that it is not the cause of death), perimortem (occurring around the time of death and therefore could have been the cause of death), or postmortem (enough after death that it was not the cause of death). Therefore, recognizing perimortem trauma becomes paramount in these investigations. Additionally, the determination of what caused the trauma, such as projectiles (e.g., bullets), clubs, knives (or other sharp objects) or something else becomes important. Lastly, the estimation of the number of bones injured, the direction from which the causative force hit the bones, and any other information on trauma may be important in the forensic investigation.

In some cases, a match with a missing persons' file is not found and the police ask for the public's help in identifying the remains. To do this, an approximation of the face on the skull is shown along with the biological profile of the dead person. In the past, slabs of clay were applied to the skull face on various locations using thicknesses known from studies of live or dead persons. Now reproductions are produced virtually on a 3-D model of the skull using these known thicknesses. The result of this work is a 'plain' face that may jog the memory of someone who has not seen a friend or family member for some time and is wondering what has happened to them.

A last set of methods used in forensic investigations involve determining the person represented by the skeleton. This is called making an identification and can be either a positive identification or probable identification. A positive identification, sometimes called personal identification, means that the skeleton is identified as one person to the exclusion of all other persons. There are a number of methods used for making this type of identification, the oldest of which is the shape and size of the frontal sinuses as seen in radiographs (x-rays). These structures are considered unique to individuals and are compared between those seen in the skeleton of interest and those of radiograph taken of a person while alive. If the two match in size and shape, a positive identification is made. Other methods use the pattern of trabecular bone in various parts of the skeleton to match between the case and a radiograph (x-ray) taken of a person in life. For example, using an x-ray of the mastoid process of an unknown case, the trabecular pattern (as well as general outline of the process) can be matched with an x-ray of possible match. If the patterns match when one is placed over the other, a positive identification is made. Other methods use any unique form of a bone or repaired bone to make a match. In one case, an imperfectly healed break of a femur had such a unique outline that a match with an x-ray was considered a positive identification. In another case, a positive identification was made by matching an odd shape of a collar bone (clavicle) with an x-ray of a person who was the same sex, age, and stature. In addition to these, another source of positive identification is surgical and dental implants that have manufacturer's numbers stamped into them. These numbers can be matched to the purchasing doctor or dentist, who then can make a positive identification.

A probable identification is one where it is highly probable that the skeleton is that of a missing person. One way this can be done is photographic superimposition. In this technique, the photographed face of a suspected match is projected onto the skull face, and points of matching and nonmatching are observed. Thus, if the eyes do not match in width and height with the eye orbits and the mouth is not located over the teeth of the skull, then the suspected missing person is excluded. However, if these features along with placement of the nose and ears do match, then a probable identification is made. Although this and similar 'photograph to skull' comparisons cannot lead to a positive identification, it can provide authorities with enough evidence to look further into the missing persons medical records for more conclusive proof of a positive identification.

## LIFEWAYS OF PAST POPULATIONS

Researchers interested in the lifeways of past people do all of the analyses described in earlier chapters of this book but combine their findings to reveal as much as possible about the lives of past people. These skeletal biology studies usually involve the remains from prehistoric and historic (formal and informal) cemeteries, where dozens, if not hundreds of skeletons are analyzed. By combining the information on ancestry, sex, and age at death, population age structure, life expectancy, and other similar information can be generated. Combined pathological data can provide information on dietary deficiencies, or disease loads. Other aggregate data may illuminate how they practiced medicine (i.e., treated diseases) or engaged in warfare. These studies use a set of methods not described in previous chapters to learn as much as possible about an extinct population from the remains of their dead.

One combination that is commonly developed are statistics (frequencies) of the sexes and ages at death of the people represented in the skeletal collection. This is the subject of paleodemography (demography of past populations), where various statistics (e.g., frequency of females in the 20 to 30 year old age bracket, frequency of males to reach 50 years of age or older) are calculated to understand key features of a past society. One of the most common tools for showing these key features is the life table. In this table, the left-hand column lists age-at-death categories (e.g., 0–5 yrs, 6–10 yrs, 11–15 yrs., etc.) for males or females, followed by columns showing the frequency of people surviving in the age category, frequency of deaths per age category, probability of death, average number of years remaining in each age category (life expectancy) and others. These calculations show how long people lived, the ages when they were most likely to die, and other such information. Comparisons of life tables between different skeletal collections may show differences in life expectancies for societies with different ways of living (e.g., hunter-gatherers vs. agriculturalists).

When reconstructing lifeways of past peoples, the relationship between the people from an osteological collection being studied and other collections in the same area or even worldwide are also of interest. These studies of 'biological distance' use the frequencies of nonmetric traits described in Chapter 7 to help understand the movement of populations through space and time. To better use these frequencies, they are combined into statistics called measures of divergence or measures of distance, whose size shows the closeness of populations. (As a reminder, differences in frequencies between groups are due to genetic drift that occurs when populations get so far apart that interbreeding is less possible if not impossible.) As an example, in a worldwide study of the relationship between a number of different populations, the mean measure of divergence (MMD) between a Palestinian sample and an Egyptian sample was MMD=.045 while the same statistic between Egyptian and Native American skulls from British Columbia (Canada) was MMMD=.143. The larger MMD indicates greater genetic distance (genetic divergence) between Egyptians and the Native Americans than between Egyptians and Palestinians. This is expected since the two Middle Eastern countries are geographically so much closer to each other than they are to North America.

Not only can frequency differences help to distinguish related from unrelated populations, but they can also show differences between groups within a population. In one landmark study, it was discovered that male skeletons in different villages had very similar frequencies of traits while the frequencies of the female skeletons were much more different. Since other studies have shown in the nonmetric traits used in this analysis are not more common in one sex over the other (i.e., they are not sex-linked), it appears that the women came from different populations than the men, which in turn indicated a patrilocal post-marital residence (i.e., women came from different villages and lived with their husband's family). Other studies show that nonmetric trait frequencies, when used with a measure of divergence, such as the mean measure of divergence, can show types of post-martial residence.

Another feature that is analyzed in lifeway studies is the internal and external shape and structure of bones to get clues on how they changed with different stresses caused by different activities. Since bone is living tissue, it can change its shape when a person does a task repeatedly, especially if it is a strenuous task (see Chapter 7 "Skeletal Anomalies"). Sometimes the bone simply gets thicker, such as the increase in cortical bone thickness observed in the humerus of the arm used by professional tennis players (as compared to the other side). A favored analysis is the cross-sectional shape of the shafts of long bones, especially the femur, which helps researchers understand the types of forces individuals put on their bodies (muscles, tendons, and bones) which in turn helps to understand the types of activities these individuals performed. For example, the front-to-back (anterior-posterior) and side-to-side (medial-lateral) measurements of the femoral shaft are almost equal in prehistoric farmers indicating that the shafts are nearly circular while the femoral shafts of hunter-gatherers are more flattened side-to-side (mediolaterally). These findings are worldwide in that flattened femoral shafts are common in hunter-gatherers while circular shafts are more common in farmers. The constant moving of hunter-gatherers in search of food appears to put more front-to-back (anterior-posterior) forces on the leg bones than does the more sedentary life of farmers. The same thing can be seen in males versus females in all peoples no matter what the subsistence pattern: the femurs of males are more flattened mediolaterally than females, apparently due to the more physically demanding work of males. These and other, more complex measurements help to untangle the stresses on the bones of past peoples, which leads to a better understanding of their day-to-day activities.

Another analysis that helps untangle parts of lifeways are stress makers in bone and teeth that show when various stresses occurred during growth. As described in Chapter 8, many people in earlier times show horizontal bands of thin enamel on their teeth, especially the incisors that appear like ripples in the otherwise smooth labial surface. The location of these so-called hypoplasias shows at what year during growth children suffered from malnutrition or other stressor (see Chapter 5 "Estimating Age at Death"). For example, looking at Figure 5.1, about half of the crown of the upper central incisor is formed (in the crypt) by around 3 years of age. Thus, if a 'ripple' appears halfway down the crown of a central incisor, the person suffered some fairly severe stressor at that age. Another indicator of stress is porotic hyperostosis described in Chapter 8. As noted, this disorder is mainly due to iron deficiency (anemia) and causes the pores seen in Figure 8.2D to appear in the otherwise smooth bone of the cranial vault and upper eye orbits. Since age of subadults can be told from their teeth and other sources, the age by which these pores appear can be told. This again helps to estimate when children experienced stress. A third indicator of stress are thin sheets of bone across the medullary cavity in long limb bones, especially the tibia. Called Harris lines, these lenses form at the location of the line between the diaphysis and epiphysis when growth stops for a time due to some stressor. Visible as lines in x-rays, the location of these lenses indicates the age at which the stressor occurred since the percentages of growth in long bones at various ages are fairly well known. This also helps to show the age at which children encounter stress. The frequencies of all of these stress markers and the ages when they occurred in male and female skeletons may show differences in the treatment of the sexes, revealing more about societal customs. Similarly, comparing the frequencies of stress markers of hunter-gatherers and farmers can lead to a better understanding of the different dietary and other stressors faced by people who collected foods occurring in nature versus those who grew their food.

Another special method involves detecting the presence of stable (non-radioactive) isotopes and other elements in bones to better understand the diet, nutrition, and life history of the people represented by a skeletal collection. The stable carbon isotopes found in the bone are particularly useful for revealing the likely diet of people. Diets that do not include large amounts of maize leave a different amount of caron isotope than diets that do include maize. As described in Chapter 1, a study of many skeletons from prehistoric Ontario showed the amount of the maize isotope was small and fairly stable from 3000 BC to around AD 500, after which it increased dramatically. This indicated that the people ate more maize from this later time up until around AD 1500, when they presumably became full-time maize farmers. In a study of prehistoric people in the Great Salt Lake region, there was a diversity of carbon isotopes in the skeletons from AD 400 to around AD 1000, after which diversity decreased indicating that the people from that time onward ate a less diverse diet. These and other isotopes and other elements reveal much information on the diets of past peoples.

Unique to studies that untangle the lifeways of past populations is the calculation of the number of people represented in a collection of skeletons from either formal or informal cemeteries. Unless it comes from an undisturbed formal cemetery with coffins, most collections have incomplete skeletons along with full skeletons. There are usually a large number of single bones or partial skeletons that cannot be easily associated with other bones and assembled into a complete, or nearly complete, skeleton. This makes it difficult for researchers to get an accurate count of how many individuals are in the collection since the number of full skeletons would not include those persons represented by partial skeletons or single bones. In these cases, the most common unique bone is used as a count of the number of people in the collection. This results in what is called the minimum number of individuals (MNI). For example, if 20 left humeri and 35 right femurs are found, both within (nearly) full skeletons and as individual bones, then the number of people represented in the collection is at least 35. Another calculation is what's called the most likely number of individuals (MLNI). Since it is possible that some of the 20 left humeri in the above example do not belong to the 35 right femurs, the formula for MLNI compensates for this situation by using the highest number of right and left bones, both individual and from a single skeleton, present in the collection. As an example, suppose a collection of bones has 18 left humeri and 20 right humeri of which 6 are paired (i.e., both sides are present from 6 individuals). Where MNI in this collection is 20, the calculation of MLNI (which is too complex for a book of this nature) gives 56. Research has shown that MNI gives a good estimate of the number of people in a collection if most are represented by full, or nearly full, skeletons. However, when there are few full skeletons but many individual bones, then MLNI gives the best estimate.

## DISEASE IN THE PAST

Some skeletal biologists study diseases in past populations and focus on the identification of pathological conditions and their effects on the lives of people. In much the same way that studies of lifeways of extinct populations aggregate data, these specialists combine pathological information from multiple skeletons to reveal the origin of diseases and their spread through time and/or between towns, countries and even continents. Their studies may also include research on how sex differences in disease frequency may infer cultural traditions and even gender roles. These and other researches related to pathological conditions use the presence of disease to understand the disease load of populations represented by their skeletons.

The most important task of paleopathology is identifying (diagnosing) diseases (pathological conditions) in bone. Many books and articles describe the way different diseases appear in bone so that these conditions can be more accurately identified. However, since different diseases have similar appearances in bone, diseases are often categorized into one of many basic types. In this book, a shortened number of basic types was described in Chapter 8 in keeping with the complexity of a book of this nature. However, full books on paleopathology have a large number of basic types such as infectious diseases, bacterial infections, infections due to fungi, viruses, and parasites, circulatory disturbances, metabolic disease, endocrine disorders, trauma, and many others. Placing a diseased bone into one of these types is often the first step in diagnosing a pathological condition. After this step, a more specific diagnosis is usually attempted (but is often not possible).

Another method specific to paleopathology is the identification of the disease-causing bacteria or viruses (collectively called pathogens). These microorganisms are identified by their DNA extracted from the bone and teeth of past people and have several uses. First, their presence helps to confirm a diagnosis of a particular disease. For example, some diseases cause bone changes that are similar to others such as syphilis and leprosy which both cause destruction of the area around the nose. Thus, for cases with this type of nasal destruction, the presence of DNA of one of the pathogens would make it likely that it was the cause of the disease. Second, the presence of pathogen DNA can be used to identify where the bacteria or virus came from. For example, in the 1500s of what is now Mexico, a large number of native Americans died after the European conquest. This 'die off' was originally thought to be due to pathogens that Europeans brought with them that killed the local people due to their lack of natural immunity. However, when DNA was extracted from the skeletons of native Americas in a cemetery from this time period, researchers found the presence of pathogens that originated in Africa. Thus, a large part of the die off was caused by pathogens brought in by African slaves. Third, pathogen DNA can be used to trace the history of disease through time and space. For example, a study of 1867 skeletons from Eurasia and the Americas revealed the presence of the pathogen for smallpox as early as AD 600 in northern Europe and Western Russia, which is over 1,000 years earlier than the oldest known case up to that point. Fourth, DNA can be used to identify the presence of a disease which rarely causes bone changes. Such a disease is smallpox that has killed millions of people throughout human history. Presence of the DNA of this pathogen (the variola virus) shows that the disease was present even when there are no skeletal changes.

Researchers who study diseases often use imaging technology, such as x-rays, CT-scans, scanning electron microscopy (SEM), and magnetic resonance imaging (MRI), when trying to identify diseases in a skeletal collection. While most pathological conditions with bone changes can be seen with the unaided eye, some cause internal bone changes that can only be seen by these technologies. The Harris Lines described above are a good example, as without x-rays (or similar technologies), their presence would not be known, and a major indicator of ill health would be missed.

Histology is another area of study that helps uncover ancient diseases. This involves sectioning bone into thin slices that can be stained to help illuminate its internal microscopic structure and then examining under a microscope. The main features of interest are the density of complete and fragmentary osteons, and the rate of bone formation. These factors are known in healthy people, so if a skeleton is encountered with different densities and formation rates, it is reasonably certain that the person represented by the bones suffered from some sort of disease. This expectation was validated when the density of osteons and bone formation rate was found to be greater in a person suffering from metastatic breast cancer than expected from a healthy person. Thus, ancient bones with similar histological density and formations probably indicates the presence of some a disease.

## EVOLUTION OF FOSSIL PREHUMANS AND HUMANS

In addition to studying modern humans, skeletal biologists also study early humans and their ancestors (prehumans). These osteologists excavate in areas where they hope to find fossil ancestors and then analyze their finds. Their goals are similar to those of other skeletal biologists; that is, an understanding of the way past humans and prehumans lived. In their pursuit of this goal, they make the same observations and measurements described in the earlier chapters in this book, but also look at other things. They look to see if their fossil finds fit into a known species, such as *Australopithecus africanus* or *Homo erectus*, which helps to reveal the amount of variation in these species. However, if their fossil does not share enough traits with a known species or it has characteristics from two different species, then they may give it a different name and/or place it as a transitional form between earlier and later human ancestors. This is a different version of estimating ancestry and is part of a larger goal of reconstructing the evolution of modern humans from earlier (prehuman) forms.

Research into fossil humans and prehumans use information on bone form and function to determine if the individual walked upright, was able to use their hands to manipulate objects (e.g., use tools), or other information on behavior. In the pursuit of these goals, a number of skeletal features are tracked that the previously described specialists generally do not. One such feature is the size of the brain. The human brain is one of the largest in the world and is tied to the many innovations and complex societies that have developed throughout human prehistory and history. Since the brain has increased in size dramatically over the last millions of years, its size at various points in the evolutionary tree is of great interest. However, the brain itself is rarely preserved so the size of the inside of the braincase (endocranial volume) is used as a substitute for brain size. Modern humans have an endocranial volume of around 1400 cubic centimeters (cc's) with females averaging around 1350 ccs while males are around 1450. Early prehumans from 3.5 million years ago (MYA), had an average volume of 500 ccs, meaning that brain size increased almost three times that amount during human evolution.

Measuring endocranial volume requires methods not regularly used by other skeletal biologists. If a braincase is empty (i.e., does not contain parts of the brain and its surrounding tissues), it can be filled with small metal balls, such as bb's for a bb gun, which can then be emptied into a graduated cylinder that measures the amount in ccs. This technique is useful for modern human skulls and some fossil crania, but many times the inside of the braincase is not empty and imaging technology must be used to generate a 3-D model that can be used to calculate cranial capacity. Whatever technique is used, estimating this trait in prehumans and early humans is one of the major tasks of the study of fossil ancestors.

Another feature that is tracked is the placement of the foramen magnum through time. In modern humans, this opening is centrally located under the braincase so that the skull is balanced on the vertebral column. In prehumans, this foramen is located more to the back of the skull, and in really early ancestors it can be located at the angle between the bottom of the skull and the back of the skull. This forward progression helps to track when early human ancestors began to walk upright (became bipedal).

The shape of the bones of the pelvis also aids in the estimation of when bipedalism became more common. The backwardly jutting structure of the ilium is the most visible shape that offers a clue to habitual walking on two legs. This backward jut provides a lever arm for the muscles that attach the posterior ilium (gluteus maximus) to the posterior femur and hold the pelvis level and at right angles to the legs (which is necessary for bipedalism). This and other pelvic and leg features of fossil ancestors are intensely studied.

The above description of the pelvis and bipedalism is part of the larger study of the relationship between function and form, called functional morphology (morphology is the study of form and structure). Persons who study fossil humans and prehumans must be able to recognize the relationship between the form of bones and any activities (functions) that they were used to perform. For example, the glenoid cavity of modern humans is deeper (concave) and more rounded than earlier prehumans such as *Australopithecus*. Since throwing projectile weapons (e.g., spears) has been shown to put stresses on the glenoid cavity (as well as the rest of the shoulder bones), this morphology of the cavity in modern humans may reflect the need to compensate for the stresses of object throwing, an action that earlier forms may not have performed. Analyses such as this with other bones are carried out on the various pre-humans and modern humans in all places and through all times.

## SUMMARY

1   In addition to the methods discussed in this book, there are a variety of others that various specialists in human skeletal biology use in their work.
2   There are four such areas of study that use special methods: forensic investigations, extinct population lifeways, paleopathology, and fossil humans and prehumans.
3   In a forensic investigation, skeletal biologists use special methods to help identify a person from their bones as well as get any information on how they died.
4   Osteologists who study the lifeways of extinct populations sum the data on all persons in a skeletal collection to illuminate their lives and how they lived.
5   Paleopathologists look at pathological conditions in the bones of people to help understand the disease load and other features of their lives.
6   The analysis of fossil humans or their ancestors involves exploring features of bone useful for understanding how humans evolved.

## FURTHER READING

For forensic investigations: SN Byers and CA Juarez, *Introduction to Forensic Anthropology*, 6th ed. (New York, NY: Routledge, 2023), and AM Christensen, NV Passalacqua, and EJ Bartelink, *Forensic Anthropology: Current*

*Methods and Practice*, 2nd ed. (Cambridge, MA: Academic Press, 2019). For extinct population lifeways: CS Larsen, *Bioarchaeology: Interpreting Behavior from the Human Skeleton*, 2nd ed. (Cambridge, UK: Cambridge University Press, 2015). For paleopathology: JA Buikstra, ed., *Ortner's Identification of Pathological Conditions in Human Skeletal Remains*, 3rd ed. (Cambridge, MA: Academic Press, 2019). For fossil humans and prehumans: FH Smith, M Cartmill, *The Human Lineage*, 2nd ed. (Hoboken, NJ: Wiley-Blackwell, 2022).

# 11

# CONCLUSION

As mentioned in Chapter 1, the purpose of this book is to present the reader with the basic observations and metric measurements used to reveal features of the lives of people represented only by their skeletons. The author hopes that this book has helped you, the reader, have a better understanding of what the human skeleton says about the life, and sometimes death, of a person. After learning about the bones and teeth that make up the human body, the reader should have been able to see how the different features, bone shapes and measurements help to reveal parts of a person's life such as their ancestry, sex, age at death, and living height. Similarly, the common anomalies and pathological conditions as well as cultural modifications that are visibly different from what is considered 'normal' for the human skeleton tell other things about a person's life. Taken together, all of the subjects discussed in this book show that bones reveal much more about a person than most people probably imagine.

Although not explicitly stated, the human skeletal biology described in this book is a subfield of the much larger field of anthropology. Anthropology is the study of humans in all places and through all of time. To break down this imposing subject, the field is divided into four subfields, each of which examines a different (with some overlap) part of this broad subject. One subfield, socio-cultural anthropology, is devoted to the study of human behavior in living or recently extinct cultures. Persons in this specialty describe and analyze what people from different areas of the world do/did to survive and reproduce. Their descriptions usually include revealing how they get their food, how they organize their

DOI: 10.4324/9781003487944-11

society (families, tribes, chiefdoms, kingdoms), what is their religion, what music they play and arts they create, and many other parts of their society. Archaeology, another subfield, is the study of past cultures from their tools, dwellings, food remains, and any object made and/or used by humans in the past. People in this specialty mostly excavate prehistoric and historic sites of (usually) extinct cultures to try to reconstruct the same features of society as socio-cultural anthropologists; that is, how they got their food, how they arranged their society, what was their religion, and etc. A third subfield is linguistic anthropology, which is the study language, and especially its association to culture. Specialists in this subfield explore the relationship between different languages, their basic structure, and how language and thought interact. The last subfield is biological anthropology, formerly known as physical anthropology. Practitioners of this subfield study the biological evolution of humans, and the variation seen in modern humans resulting from that evolution. Although it has a number of different specialties like population genetics and paleo-genetics as well as primate behavior, the majority of these specialists study bones to answer questions about the lives of people represented only by their skeletal remains. Their work is characterized not only by the subjects described in Chapters 2 through 9 of this book but also by the methods described in Chapter 10. All involve the study of bones.

Biological anthropology is unique from all other scientific disciplines in that it is the study of the bones and teeth of dead people with the intention of understanding the lives of these people. This makes it different from health care and other professions that deal with the human body. Physicians also look at bones, but they are usually only interested in those of living people, such as which are broken and need mending, or which have pathological conditions that need to be treated. Similarly, nurses study the body to help in their work of facilitating the healing of patients. Anatomists also look at bones, but it is a small part of their study of all parts of human beings, which includes internal organs, nerves, muscles, and all other soft tissues. Dentists study human teeth so they can repair damaged or decayed teeth and restore oral health. Only biological anthropologists study skeletal and dental remains to see what they tell us about the lives of people.

As with anthropology in general, biological anthropology is divided into subspecialties. The final chapter of this book discussed some of these by describing special methods used by four different types of biological anthropologists in their analyses of human skeletons. Those bone specialists who work on forensic cases are called forensic anthropologists, while those who work to understand past populations from the summed data of skeletal collections are called bioarchaeologists. Those who use the remaining two special methods also have special names; paleopathologists study ancient diseases and their effects on populations, and those who study fossil humans and prehumans are called paleoanthropologists. All are subspecialists of the subfield of biological anthropology.

As stated in Chapter 1, the reader is not an expert human skeletal biologist after reading this book. As mentioned above, such expertise is only gained by intense study over a lifetime of viewing osteological remains. Studying human bone is both an exhilarating and frustrating activity in that so much can be estimated from them, but so much more remains unknown. Although the osteological traits that help identify parts of people's lives seem fairly straightforward, the reader must remember that all of these are best thought of as probabilities. What is predicted about the life of persons from their skeletons can be in error and never should be considered hard-and-fast. However, the author hopes the reader understands some of what can be told (or, more appropriately, 'estimated') about the life and (sometimes) death of persons represented only by their skeleton.

# INDEX

Note: Page numbers in *italics* refer to figures and those in **bold** to tables.

For Product Safety Concerns and Information please contact our EU
representative GPSR@taylorandfrancis.com
Taylor & Francis Verlag GmbH, Kaufingerstraße 24, 80331 München, Germany

www.ingramcontent.com/pod-product-compliance
Lightning Source LLC
Chambersburg PA
CBHW050651270326
41927CB00012B/2974